Physical Science

STUDY GUIDE

D1088898

This book was printed with soy-based ink on acid-free recycled content paper, containing 10% POSTCONSUMER WASTE.

HOLT, RINEHART AND WINSTON

A Harcourt Classroom Education Company

Austin • New York • Orlando • Atlanta • San Francisco • Boston • Dallas • Toronto • London

To the Student

Are you looking for a way to keep all of your vocabulary terms straight? Do you need practice for an upcoming test? If so, then this booklet is for you. The *Study Guide* is a tool that allows you to confirm what you know and identify topics of difficulty, so that you can succeed in your study of physical science. These worksheets are reproductions of the Chapter Highlights and Chapter Review sections that follow each chapter in the textbook, with one difference—the *Study Guide* worksheets provide plenty of space for you to record your answers and write down your thoughts and ideas.

■ VOCABULARY & NOTES WORKSHEETS

Vocabulary & Notes Worksheets serve as an important tool to help you organize what you have learned from the chapter. You can use these worksheets

- as a reading guide, to help you identify and study the main concepts of each chapter before or after you read each section;
- as a place to record and review the definitions of important vocabulary terms from each chapter;
- as a reference, to help you study for exams and determine which topics you have learned well and which topics you may need to study further.

■ CHAPTER REVIEW WORKSHEETS

Chapter Review Worksheets give you a chance to practice using what you have a learned from the chapter. You can use these worksheets

- as a learning tool, to work interactively with the textbook by answering the questions that relate to the chapter as you read the text;
- as a review, to test your understanding of the chapter's main concepts and terminology;
- as a practice test, to prepare you for taking the Chapter Test.

Art and Photo Credits
All work, unless otherwise noted, contributed by Holt, Rinehart and Winston.
Abbreviated as follows: (t) top; (b) bottom; (l) left; (r) right; (c) center; (bkgd) background.
Front cover (owl), Kim Taylor/Bruce Coleman, Inc.; (bridge), Henry K. Kaiser/Leo de Wys; (dove), Stephen Dalton/Photo Researchers, Inc.; Page 19 (tc), HRW Artist [photo reference: Lance Schriner/HRW Photo]; 26 (c), HRW Artist [photo reference: Charles D. Winters/Photo Researchers, Inc.]; 28 (tc), Preface, Inc.; 37 (c), Preface, Inc.; 38 (tc), Preface, Inc.; 46 (bc), Preface, Inc.; 47 (c), Preface, Inc.; 55 (tc), HRW Artist [photo reference: NASA]; 63 (b), Jared Schneidman/Wilkinson Studios; 72 (c), HRW Artist [photo reference: John Langford/HRW Photo; 72 (b), HRW Artist [photo reference: Stephanie Morris/HRW Photo]; 81 (b), David Chapman; 82 (c), Dave Joly; 92 (t), Preface, Inc.; 99 (c), Stephen Durke/Washington Artists; 100 (c), Stephen Durke/Washington Artists; 107 (c), Annie Bissett; 115 (c), HRW Artist [photo reference: Sam Dudgeon/HRW Photo]; 116 (c), HRW Artist [photo reference: Sam Dudgeon/HRW Photo]; 126 (c), Preface, Inc.; 143 (c), Preface, Inc.; 153 (c), HRW Artist [photo reference: John Langford/HRW Photo; 161 (c), Stephen Durke/Washington Artists; 175 (c), Tom Gagliano; 178 (c), Sidney Jablonski; 187 (c), Sidney Jablonski; 203 (c), Mark Heine

Printed in the United States of America

ISBN 0-03-054399-1   4 5 6 7   085   04 03 02

▪ CONTENTS ▪

CONTENTS, CONTINUED

Name _____ Date _____ Class _____

The World of Physical Science

By studying the Vocabulary and Notes listed for each section below, you can gain a better understanding of this chapter.

SECTION 1

Vocabulary

In your own words, write a definition of the following term in the space provided.

1. physical science _____

_____

Notes

Read the following section highlights. Then, in your own words, write the highlights in your ScienceLog.

• Science is a process of making observations and asking questions about those observations.

• Physical science is the study of matter and energy and is often divided into physics and chemistry.

• Physical science is part of many other areas of science.

• Many different careers involve physical science.

SECTION 2

Vocabulary

In your own words, write a definition of each of the following terms in the space provided.

1. scientific method _____

_____

_____

2. technology _____

_____

_____

3. observation _____

_____

4. hypothesis _____

_____

5. data _____

_____

6. theory _____

_____

_____

7. law _____

_____

Notes

Read the following section highlights. Then, in your own words, write the highlights in your ScienceLog.

• The scientific method is a series of steps that scientists use to answer questions and solve problems.

• Any information you gather through your senses is an observation. Observations often lead to questions or problems.

• A hypothesis is a possible explanation or answer to a question. A good hypothesis is testable.

• After you test a hypothesis, you should analyze your results and draw conclusions about whether your hypothesis was supported.

• Communicating your findings allows others to verify your results or continue to investigate your problem.

• A scientific theory is the result of many investigations and many hypotheses. Theories can be changed or modified by new evidence.

• A scientific law is a summary of many experimental results and hypotheses that have been supported over time.

SECTION 3

Vocabulary

In your own words, write a definition of the following term in the space provided.

1. model _____

_____

Notes

Read the following section highlights. Then, in your own words, write the highlights in your ScienceLog.

• Scientific models are representations of objects or systems. Models make difficult concepts easier to understand.

• Models can represent things too small to see or too large to observe directly.

• Models can be used to test hypotheses and illustrate theories.

SECTION 4

Vocabulary

In your own words, write a definition for each of the following terms in the space provided.

1. meter _____

_____

_____

_____

_____

_____

2. volume _____

_____

_____

_____

_____

_____

3. mass _____

_____

_____

_____

_____

_____

4. temperature _____

_____

_____

_____

_____

5. area _____

_____

_____

_____

_____

6. density _____

_____

_____

_____

_____

Notes

Read the following section highlights. Then, in your own words, write the highlights in your ScienceLog.

- The International System of Units is the standard system of measurement used by scientists around the world.
- Length, volume, mass, and temperature are quantities of measurement. Each quantity of measurement is expressed with a particular SI unit.
- Area is a measure of how much surface an object has. Density is a measure of mass per unit volume.
- Safety rules are important and must be followed at all times during scientific investigations.

Name _____ Date _____ Class_____

The World of Physical Science

USING VOCABULARY

For each pair of terms, explain the difference in their meanings.

1. science/technology _____

_____

_____

_____

2. observation/hypothesis _____

_____

_____

_____

3. theory/law _____

_____

_____

_____

4. model/theory _____

_____

_____

_____

5. volume/mass _____

_____

_____

_____

6. area/density _____

_____

_____

_____

UNDERSTANDING CONCEPTS

Multiple Choice

7. Physical science is the study of

   a. matter and motion.

   b. matter and energy.

   c. energy and motion.

   d. matter and composition.

8. 10 m is equal to

   a. 100 cm.

   b. 1,000 cm.

   c. 10,000 mm.

   d. Both (b) and (c)

9. For a hypothesis to be valid, it must be

   a. testable.

   b. supported by evidence.

   c. made into a law.

   d. Both (a) and (b)

10. The statement "Sheila has a stain on her shirt" is an example of a(n)

   a. law.

   b. hypothesis.

   c. observation.

   d. prediction.

11. A hypothesis is often developed out of

   a. observations.

   b. experiments.

   c. laws.

   d. Both (a) and (b)

12. How many milliliters are in 3.5 kL?

   a. 3,500

   b. 0.0035

   c. 3,500,000

   d. 35,000

13. A map of Seattle is an example of a

   a. law.

   b. quantity.

   c. model.

   d. unit.

14. Which of the following is an example of technology?

   a. mass

   b. physical science

   c. screwdriver

   d. None of the above

CHAPTER 1

Short Answer

15. Name two areas of science other than chemistry and physics, and describe how physical science has a role in those areas of science.

_____

_____

_____

_____

_____

16. Explain why the results of one experiment are never really final results.

_____

_____

_____

_____

_____

17. Explain why area and density are called derived quantities.

_____

_____

_____

_____

_____

18. If a hypothesis is not testable, does that mean that it is wrong? Explain.

_____

_____

_____

_____

_____

CONCEPT MAPPING

19. Use the following terms to create a concept map: *science, scientific method, hypothesis, problems, questions, experiments, observations.*

The World of Physical Science, continued

CRITICAL THINKING AND PROBLEM SOLVING

Write one or two sentences to answer each of the following questions.

20. A tailor is someone who makes or alters items of clothing. Why might a standard system of measurement be helpful to a tailor?

_____

_____

_____

21. Two classmates are having a debate about whether a spatula is an example of technology. Using what you know about science, technology, and spatulas, write a couple of sentences that will help your classmates settle their debate.

_____

_____

_____

22. Imagine that you are conducting an experiment in which you are testing the effects of the height of a ramp on the speed at which a toy car goes down the ramp. What is the variable in this experiment? What factors must be controlled?

_____

_____

_____

23. Suppose a classmate says, "I don't need to study physical science because I'm not going to be a scientist, and scientists are the only people who use physical science." How would you respond? (Hint: In your answer, give several examples of careers that use physical science.)

_____

_____

_____

MATH IN SCIENCE

24. A cereal box has a mass of 340 g. Its dimensions are 27 cm $\times$ 19 cm $\times$ 6 cm.

a. What is the volume of the box?

_____

_____

The World of Physical Science, continued

b. What is its density?

_____

_____

c. What is the area of the front side of the box?

_____

_____

INTERPRETING GRAPHICS

Examine the graphic on page 31 of your book and answer the following questions.

25. How similar to the real object is this model?

_____

_____

_____

_____

26. What characteristics of the real object does this model not show?

_____

_____

_____

_____

27. Why might this model be useful?

_____

_____

_____

_____

NOW WHAT DO YOU THINK?

Take a minute to review your answers to the ScienceLog questions at the beginning of this chapter. Have your answers changed? If necessary, revise your answers based on what you have learned since you began this chapter. Record your revisions in your ScienceLog.

CHAPTER

2 VOCABULARY & NOTES WORKSHEET

The Properties of Matter

By studying the Vocabulary and Notes listed for each section below, you can gain a better understanding of this chapter.

SECTION 1

Vocabulary

In your own words, write a definition for each of the following terms in the space provided.

1. matter _____

_____

_____

2. volume _____

_____

_____

_____

3. meniscus _____

_____

_____

_____

4. mass _____

_____

_____

_____

5. gravity _____

_____

_____

6. weight _____

_____

_____

CHAPTER 2

7. newton _____

_____

8. inertia _____

_____

Notes

Read the following section highlights. Then, in your own words, write the highlights in your ScienceLog.

- Matter is anything that has volume and mass.
- Volume is the amount of space taken up by an object.
- The volume of liquids is expressed in liters and milliliters.
- The volume of solid objects is expressed in cubic units, such as cubic meters.
- Mass is the amount of matter that something is made of.
- Mass and weight are not the same thing. Weight is a measure of the gravitational force exerted on an object, usually in relation to the Earth.
- Mass is usually expressed in milligrams, grams, and kilograms.
- The newton is the SI unit of force, so weight is expressed in newtons.
- Inertia is the tendency of all objects to resist any change in motion. Mass is a measure of inertia. The more massive an object is, the greater its inertia.

SECTION 2

Vocabulary

In your own words, write a definition for each of the following terms in the space provided.

1. physical property _____

_____

_____

_____

_____

2. density _____

_____

_____

_____

_____

The Properties of Matter, continued

3. chemical property _____

_____

_____

_____

_____

4. physical change _____

_____

_____

_____

_____

5. chemical change _____

_____

_____

_____

_____

Notes

Read the following section highlights. Then, in your own words, write the highlights in your ScienceLog.

- Physical properties of matter can be observed without changing the identity of the matter.
- Density is the amount of matter in a given space, or the mass per unit volume.
- The density of a substance is always the same at a given pressure and temperature regardless of the size of the sample of the substance.
- Chemical properties describe a substance based on its ability to change into a new substance with different properties.
- Chemical properties can be observed only when one substance might become a new substance.
- The characteristic properties of a substance are always the same whether the sample observed is large or small.
- When a substance undergoes a physical change, its identity remains the same.
- A chemical change occurs when one or more substances are changed into new substances with different properties.

The Properties of Matter

USING VOCABULARY

For each pair of terms, explain the difference in their meanings.

1. mass/volume _____

_____

_____

2. mass/weight _____

_____

_____

_____

3. inertia/mass _____

_____

4. volume/density _____

_____

_____

_____

5. physical property/chemical property _____

_____

_____

_____

_____

6. physical change/chemical change _____

_____

_____

_____

_____

UNDERSTANDING CONCEPTS

Multiple Choice

7. Which of these is NOT matter?

    a. a cloud   **c.** sunshine

    b. your hair   **d.** the sun

8. The mass of an elephant on the moon would be

    a. less than its mass on Mars.

    b. more than its mass on Mars.

    c. the same as its weight on the moon.

    d. None of the above

9. Which of the following is NOT a chemical property?

    a. reactivity with oxygen

    b. malleability

    c. flammability

    d. reactivity with acid

10. Your weight could be expressed in which of the following units?

    a. pounds

    b. newtons

    c. kilograms

    d. Both (a) and (b)

11. You accidentally break your pencil in half. This is an example of

    a. a physical change.

    b. a chemical change.

    c. density.

    d. volume.

12. Which of the following statements about density is true?

    a. Density depends on mass and volume.

    b. Density is weight per unit volume.

    c. Density is measured in milliliters.

    d. Density is a chemical property.

13. Which of the following pairs of objects would have the greatest attraction toward each other due to gravity?

    a. a 10 kg object and a 10 kg object, 4 m apart

    b. a 5 kg object and a 5 kg object, 4 m apart

    c. a 10 kg object and a 10 kg object, 2 m apart

    d. a 5 kg object and a 5 kg object, 2 m apart

14. Inertia increases as _____ increases.

    a. time   **c.** mass

    b. length   **d.** volume

CHAPTER 2

The Properties of Matter, continued

Short Answer

15. In one or two sentences, explain the different processes in measuring the volume of a liquid and measuring the volume of a solid.

_____

_____

_____

_____

_____

_____

16. In one or two sentences, explain the relationship between mass and inertia.

_____

_____

_____

_____

17. What is the formula for calculating density?

_____

_____

_____

18. List three characteristic properties of matter.

_____

_____

_____

_____

CONCEPT MAPPING

19. Use the following terms to create a concept map: *matter, mass, inertia, volume, milliliters, cubic centimeters, weight, gravity.*

CRITICAL THINKING AND PROBLEM SOLVING

20. You are making breakfast for your picky friend, Filbert. You make him scrambled eggs. He asks, "Would you please take these eggs back to the kitchen and poach them?" What scientific reason do you give Filbert for not changing his eggs?

_____

_____

_____

21. You look out your bedroom window and see your new neighbors moving in. Your neighbor bends over to pick up a small cardboard box, but he cannot lift it. What can you conclude about the item(s) in the box? Use the terms *mass* and *inertia* to explain how you came to this conclusion.

_____

_____

_____

_____

The Properties of Matter, continued

22. You may sometimes hear on the radio or on television that astronauts are "weight-less" in space. Explain why this is not true.

_____

_____

_____

_____

_____

_____

23. People commonly use the term *volume* to describe the capacity of a container. How does this definition of volume differ from the scientific definition?

_____

_____

_____

_____

_____

MATH IN SCIENCE

24. What is the volume of a book with the following dimensions: a width of 10 cm, a length that is two times the width, and a height that is half the width? Remember to express your answer in cubic units.

_____

_____

25. A jar contains 30 mL of glycerin (mass = 37.8 g) and 60 mL of corn syrup (mass = 82.8 g). Which liquid is on top? Show your work, and explain your answer.

_____

_____

_____

_____

The Properties of Matter, continued

INTERPRETING GRAPHICS

Examine the illustration below, and answer the following questions.

26. List three physical properties of this can.

_____

_____

27. Did a chemical change or a physical change cause the change in this can's appearance?

_____

_____

28. How does the density of the metal in the can compare before and after the change?

_____

_____

_____

29. Can you tell what the chemical properties of the can are just by looking at the picture? Explain.

_____

_____

_____

_____

NOW WHAT DO YOU THINK?

Take a minute to review your answers to the ScienceLog questions at the beginning of this chapter. Have your answers changed? If necessary, revise your answers based on what you have learned since you began this chapter. Record your revisions in your ScienceLog.

CHAPTER

3 VOCABULARY & NOTES WORKSHEET

States of Matter

By studying the Vocabulary and Notes listed for each section below, you can gain a better understanding of this chapter.

SECTION 1

Vocabulary

In your own words, write a definition for each of the following terms in the space provided.

1. states of matter _____

_____

2. solid _____

_____

3. liquid _____

_____

4. gas _____

_____

5. pressure _____

_____

6. Boyle's law _____

_____

_____

_____

7. Charles's law _____

_____

_____

_____

8. plasma _____

_____

_____

Notes

Read the following section highlights. Then, in your own words, write the highlights in your ScienceLog.

- The states of matter are the physical forms in which a substance can exist. The four most familiar states are solid, liquid, gas, and plasma.
- All matter is made of tiny particles called atoms and molecules that attract each other and move constantly.
- A solid has a definite shape and volume.
- A liquid has a definite volume but not a definite shape.
- A gas does not have a definite shape or volume. A gas takes the shape and volume of its container.
- Pressure is a force per unit area. Gas pressure increases as the number of collisions of gas particles increases.
- Boyle's law states that the volume of a gas increases as the pressure decreases if the temperature does not change.
- Charles's law states that the volume of a gas increases as the temperature increases if the pressure does not change.
- Plasmas are composed of particles that have broken apart. Plasmas do not have a definite shape or volume.

SECTION 2

Vocabulary

In your own words, write a definition for each of the following terms in the space provided.

1. change of state _____

_____

_____

2. melting _____

_____

_____

3. freezing _____

_____

_____

4. vaporization _____

_____

_____

5. boiling _____

_____

_____

6. evaporation _____

_____

_____

7. condensation _____

_____

_____

8. sublimation _____

_____

_____

Notes

Read the following section highlights. Then, in your own words, write the highlights in your ScienceLog.

- A change of state is the conversion of a substance from one physical form to another. All changes of state are physical changes.
- Exothermic changes release energy. Endothermic changes absorb energy.
- Melting changes a solid to a liquid. Freezing changes a liquid to a solid. The freezing point and melting point of a substance are the same temperature.
- Vaporization changes a liquid to a gas. There are two kinds of vaporization: boiling and evaporation.
- Boiling occurs throughout a liquid at the boiling point.
- Evaporation occurs at the surface of a liquid, at a temperature below the boiling point.
- Condensation changes a gas to a liquid.
- Sublimation changes a solid directly to a gas.
- Temperature does not change during a change of state.

CHAPTER

3 CHAPTER REVIEW WORKSHEET

States of Matter

USING VOCABULARY

For each pair of terms, explain the difference in their meanings.

1. solid/liquid _____

_____

_____

_____

2. Boyle's Law/Charles's Law _____

_____

_____

_____

3. evaporation/boiling _____

_____

_____

_____

4. melting/freezing _____

_____

_____

_____

UNDERSTANDING CONCEPTS

Multiple Choice

5. Which of the following best describes the particles of a liquid?

a. The particles are far apart and moving fast.
b. The particles are close together but moving past each other.
c. The particles are far apart and moving slowly.
d. The particles are closely packed and vibrate in place.

6. Boiling points and freezing points are examples of

a. chemical properties.　　　　**c.** energy.
b. physical properties.　　　　**d.** matter.

7. During which change of state do atoms or molecules become more ordered?

a. boiling　　　　**c.** melting
b. condensation　　　　**d.** sublimation

8. Which of the following describes what happens as the temperature of a gas in a balloon increases?
   a. The speed of the particles decreases.
   b. The volume of the gas increases and the speed of the particles increases.
   c. The volume decreases.
   d. The pressure decreases.

9. Dew collects on a spider web in the early morning. This is an example of
   a. condensation.
   b. evaporation.
   c. sublimation.
   d. melting.

10. Which of the following changes of state is exothermic?
    a. evaporation
    b. sublimation
    c. freezing
    d. melting

11. What happens to the volume of a gas inside a piston if the temperature does not change but the pressure is reduced?
    a. increases
    b. stays the same
    c. decreases
    d. not enough information

12. The atoms and molecules in matter
    a. are attracted to one another.
    b. are constantly moving.
    c. move faster at higher temperatures.
    d. All of the above

13. Which of the following contains plasma?
    a. dry ice
    b. steam
    c. a fire
    d. a hot iron

Short Answer

14. Explain why liquid water takes the shape of its container but an ice cube does not.

_____

_____

_____

_____

_____

_____

States of Matter, continued

15. Rank solids, liquids, and gases in order of decreasing particle speed.

_____

_____

16. Compare the density of iron in the solid, liquid, and gaseous states.

_____

_____

_____

_____

CONCEPT MAPPING

17. Use the following terms to create a concept map: *states of matter, solid, liquid, gas, plasma, changes of state, freezing, vaporization, condensation, melting.*

CHAPTER 3

CRITICAL THINKING AND PROBLEM SOLVING

18. After taking a shower, you notice that small droplets of water cover the mirror. Explain how this happens. Be sure to describe where the water comes from and the changes it goes through.

_____

_____

_____

_____

_____

_____

19. In the picture below, water is being split to form two new substances, hydrogen and oxygen. Is this a change of state? Explain your answer.

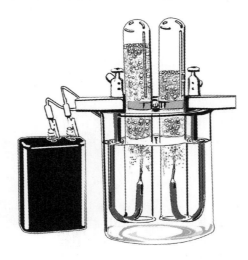

_____

_____

_____

_____

_____

_____

States of Matter, continued

---

20. To protect their crops during freezing temperatures, orange growers spray water onto the trees and allow it to freeze. In terms of energy lost and energy gained, explain why this practice protects the oranges from damage.

_____

_____

_____

_____

21. At sea level, water boils at 100°C, while methane boils at –161°C. Which of these substances has a stronger force of attraction between its particles? Explain your reasoning.

_____

_____

_____

_____

_____

MATH IN SCIENCE

22. Kate placed 100 mL of water in five different pans, placed the pans on a windowsill for a week, and measured how much water evaporated. Draw a graph of her data, shown below, with surface area on the *x*-axis. Is the graph linear or non-linear? What does this tell you?

Pan number	1	2	3	4	5
Surface area (cm²)	44	82	20	30	65
Volume evaporated (mL)	42	79	19	29	62

_____

_____

23. Examine the graph below, and answer the following questions:

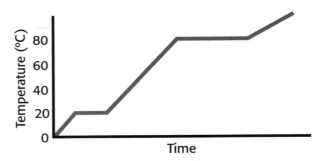

a. What is the boiling point of the substance? What is the melting point?

_____

_____

_____

b. Which state is present at 30°C?

_____

_____

_____

c. How will the substance change if energy is added to the liquid at 20°C?

_____

_____

_____

NOW WHAT DO YOU THINK?

Take a minute to review your answers to the ScienceLog questions at the beginning of this chapter. Have your answers changed? If necessary, revise your answers based on what you have learned since you began this chapter. Record your revisions in your ScienceLog.

Elements, Compounds, and Mixtures

By studying the Vocabulary and Notes listed for each section below, you can gain a better understanding of this chapter.

SECTION 1

Vocabulary

In your own words, write a definition for each of the following terms in the space provided.

1. element _____

_____

_____

_____

_____

2. pure substance _____

_____

_____

_____

3. metals _____

_____

_____

_____

4. nonmetals _____

_____

_____

_____

▲ CHAPTER 4

5. metalloids _____

_____

_____

_____

_____

Notes

Read the following section highlights. Then, in your own words, write the highlights in your ScienceLog.

- A substance in which all the particles are alike is a pure substance.
- An element is a pure substance that cannot be broken down into anything simpler by physical or chemical means.
- Each element has a unique set of physical and chemical properties.
- Elements are classified as metals, nonmetals, and metalloids based on their properties.

SECTION 2

Vocabulary

In your own words, write a definition for the following term in the space provided.

1. compound _____

_____

_____

_____

_____

Notes

Read the following section highlights. Then, in your own words, write the highlights in your ScienceLog.

- A compound is a pure substance composed of two or more elements chemically combined.
- Each compound has a unique set of physical and chemical properties that are different from the properties of the elements that compose it.
- The elements that form a compound always combine in a specific ratio according to their masses.
- Compounds can be broken down into simpler substances by chemical changes.

Elements, Compounds, and Mixtures, continued

SECTION 3

Vocabulary

In your own words, write a definition for each of the following terms in the space provided.

1. mixture _____

_____

_____

_____

2. solution _____

_____

_____

_____

3. solute _____

_____

_____

_____

4. solvent _____

_____

_____

_____

5. concentration _____

_____

_____

_____

CHAPTER 4

Elements, Compounds, and Mixtures, continued

6. solubility _____

_____

_____

_____

_____

7. suspension _____

_____

_____

_____

_____

8. colloid _____

_____

_____

_____

_____

Notes

Read the following section highlights. Then, in your own words, write the highlights in your ScienceLog.

• A mixture is a combination of two or more substances, each of which keeps its own characteristics.

• Mixtures can be separated by physical means, such as filtration and evaporation.

• The components of a mixture can be mixed in any proportion.

• A solution is a mixture that appears to be a single substance but is composed of a solute dissolved in a solvent. Solutions do not settle, cannot be filtered, and do not scatter light.

• Concentration is a measure of the amount of solute dissolved in a solvent.

• The solubility of a solute is the amount of solute needed to make a saturated solution using a given amount of solvent at a certain temperature.

• Suspensions are heterogeneous mixtures that contain particles large enough to settle out, be filtered, and block or scatter light.

• Colloids are mixtures that contain particles too small to settle out or be filtered but large enough to scatter light.

CHAPTER

4   CHAPTER REVIEW WORKSHEET

Elements, Compounds, and Mixtures

USING VOCABULARY

Complete the following sentences by writing the appropriate term from the vocabulary list in the space provided.

1. A _____ has a definite ratio of components.

2. The amount of solute needed to form a saturated solution is the

   _____ of the solute.

3. A _____ can be separated by filtration.

4. A pure substance must be either a(n) _____ or a(n)

   _____ .

5. Elements that are brittle and dull are _____ .

6. The substance that dissolves to form a solution is the _____ .

UNDERSTANDING CONCEPTS

Multiple Choice

7. Which of the following increases the solubility of a gas in a liquid?
   - **a.** increasing the temperature
   - **b.** stirring
   - **c.** decreasing the temperature
   - **d.** decreasing the amount of liquid

8. Which of the following best describes chicken noodle soup?
   - **a.** element
   - **b.** mixture
   - **c.** compound
   - **d.** solution

9. Which of the following does NOT describe elements?
   - **a.** all the particles are alike
   - **b.** can be broken down into simpler substances
   - **c.** have unique sets of properties
   - **d.** can join together to form compounds

10. A solution that contains a large amount of solute is best described as
    - **a.** unsaturated.
    - **b.** concentrated.
    - **c.** dilute.
    - **d.** weak.

11. Which of the following substances can be separated into simpler substances only by chemical means?
    - **a.** sodium   **c.** water
    - **b.** salt water   **d.** gold

▲ CHAPTER 4

Elements, Compounds, and Mixtures, continued

12. Which of the following would NOT increase the rate at which a solid dissolves?
   a. decreasing the temperature
   b. crushing the solid
   c. stirring
   d. increasing the temperature

13. An element that conducts thermal energy well and is easily shaped is a
   a. metal.
   b. metalloid.
   c. nonmetal.
   d. None of the above

14. In which classification of matter are the components chemically combined?
   a. alloy
   b. colloid
   c. compound
   d. suspension

Short Answer

15. What is the difference between an element and a compound?

_____

_____

_____

_____

_____

_____

16. When nail polish is dissolved in acetone, which substance is the solute and which is the solvent?

_____

_____

_____

_____

_____

_____

Elements, Compounds, and Mixtures, continued

CONCEPT MAPPING

17. Use the following terms to create a concept map: *matter, element, compound, mixture, solution, suspension, colloid.*

Elements, Compounds, and Mixtures, continued

CRITICAL THINKING AND PROBLEM SOLVING

18. Describe a procedure to separate a mixture of salt, finely ground pepper, and pebbles.

_____

_____

_____

_____

19. A light green powder is heated in a test tube. A gas is given off, while the solid becomes black. In which classification of matter does the green powder belong? Explain your reasoning.

_____

_____

_____

20. Why is it desirable to know the exact concentration of solutions rather than whether they are concentrated or dilute?

_____

_____

_____

21. Explain the three properties of mixtures using a fruit salad as an example.

_____

_____

_____

_____

_____

22. To keep the "fizz" in carbonated beverages after they have been opened, should you store them in a refrigerator or in a cabinet? Explain.

_____

_____

_____

_____

Elements, Compounds, and Mixtures, continued

_____

MATH IN SCIENCE

23. What is the concentration of a solution prepared by mixing 50 g of salt with 200 mL of water?

_____

_____

24. How many grams of sugar must be dissolved in 150 mL of water to make a solution with a concentration of 0.6 g/mL?

_____

_____

INTERPRETING GRAPHICS

25. Use figure below to answer the following questions:

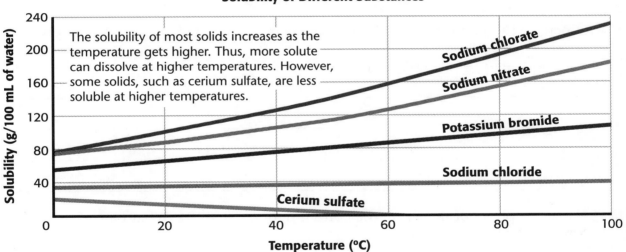

Solubility of Different Substances

The solubility of most solids increases as the temperature gets higher. Thus, more solute can dissolve at higher temperatures. However, some solids, such as cerium sulfate, are less soluble at higher temperatures.

a. Can 50 g of sodium chloride dissolve in 100 mL of water at 60°C?

_____

b. How much cerium sulfate is needed to make a saturated solution in 100 mL of water at 30°C?

_____

c. Is sodium chloride or sodium nitrate more soluble in water at 20°C?

_____

CHAPTER 4

Elements, Compounds, and Mixtures, continued

26. Dr. Sol Vent tested the solubility of a compound. The data below was collected using 100 mL of water. Graph Dr. Vent's results.

Temperature (°C)	10	25	40	60	95
Dissolved solute (g)	150	70	34	25	15

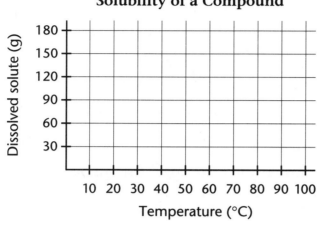

Solubility of a Compound

To increase the solubility, would you increase or decrease the temperature? Explain.

_____

_____

_____

_____

_____

27. Look at the photo of the jar on the bottom right of page 101. What type of mixture is shown in the photo? Explain.

_____

_____

_____

_____

_____

NOW WHAT DO YOU THINK?

Take a minute to review your answers to the ScienceLog questions at the beginning of this chapter. Have your answers changed? If necessary, revise your answers based on what you have learned since you began this chapter. Record your revisions in your ScienceLog.

CHAPTER

5 VOCABULARY & NOTES WORKSHEET

Matter in Motion

By studying the Vocabulary and Notes listed for each section below, you can gain a better understanding of this chapter.

SECTION 1

Vocabulary

In your own words, write a definition for each of the following terms in the space provided.

1. motion _____

_____

_____

_____

2. speed _____

_____

_____

_____

_____

_____

3. velocity _____

_____

_____

_____

_____

4. acceleration _____

_____

_____

_____

_____

Notes

Read the following section highlights. Then, in your own words, write the highlights in your ScienceLog.

- An object is in motion if it changes position over time when compared with a reference point.
- The speed of a moving object depends on the distance traveled by the object and the time taken to travel that distance.
- Speed and velocity are not the same thing. Velocity is speed in a given direction.
- Acceleration is the rate at which velocity changes.
- An object can accelerate by changing speed, changing direction, or both.
- Acceleration is calculated by subtracting starting velocity from final velocity, then dividing by the time required to change velocity.

SECTION 2

Vocabulary

In your own words, write a definition for each of the following terms in the space provided.

1. force _____

_____

_____

_____

_____

2. newton (N) _____

_____

_____

_____

_____

3. net force _____

_____

_____

_____

_____

Matter in Motion, continued

Notes

Read the following section highlights. Then, in your own words, write the highlights in your ScienceLog.

- A force is a push or a pull.
- Forces are expressed in newtons.
- Force is always exerted by one object on another object.
- Net force is determined by combining forces.
- Unbalanced forces produce a change in motion. Balanced forces produce no change in motion.

SECTION 3

Vocabulary

In your own words, write a definition of the following term in the space provided.

1. friction _____

_____

_____

_____

Notes

Read the following section highlights. Then, in your own words, write the highlights in your ScienceLog.

- Friction is a force that opposes motion.
- Friction is caused by "hills and valleys" touching on the surfaces of two objects.
- The amount of friction depends on factors such as the roughness of the surfaces and the force pushing the surfaces together.
- Four kinds of friction that affect your life are sliding friction, rolling friction, fluid friction, and static friction.
- Friction can be harmful or helpful.

SECTION 4

Vocabulary

In your own words, write a definition for each of the following terms in the space provided.

1. gravity _____

_____

_____

CHAPTER 5

Matter in Motion, continued

2. weight _____

_____

_____

_____

_____

3. mass _____

_____

_____

_____

_____

Notes

Read the following section highlights. Then, in your own words, write the highlights in your ScienceLog.

• Gravity is a force of attraction between objects that is due to their masses.

• The law of universal gravitation states that all objects in the universe attract each other through gravitational force. The size of the force depends on the masses of the objects and the distance between them.

• Weight and mass are not the same. Mass is the amount of matter in an object; weight is a measure of gravitational force on an object.

CHAPTER

5 | **CHAPTER REVIEW WORKSHEET**

Matter in Motion

USING VOCABULARY

To complete the following sentences, choose the correct term from each pair of terms listed below, and write the term in the space provided.

1. _____ opposes motion between surfaces that are touching. (Friction or Gravity)

2. Forces are expressed in _____ . (newtons or mass)

3. A _____ is determined by combining forces. (net force or newton)

4. _____ is the rate at which _____ changes. (Velocity or Acceleration, velocity or acceleration)

UNDERSTANDING CONCEPTS

Multiple Choice

5. A student riding her bicycle on a straight, flat road covers one block every 7 seconds. If each block is 100 m long, she is traveling at
   a. constant speed.
   b. constant velocity.
   c. 10 m/s.
   d. Both (a) and (b)

6. Friction is a force that
   a. opposes an object's motion.
   b. does not exist when surfaces are very smooth.
   c. decreases with larger mass.
   d. All of the above

7. Rolling friction
   a. is usually less than sliding friction.
   b. makes it difficult to move objects on wheels.
   c. is usually greater than sliding friction.
   d. is the same as fluid friction.

8. If Earth's mass doubled, your weight would
   a. increase because gravity increases.
   b. decrease because gravity increases.
   c. increase because gravity decreases.
   d. not change because you are still on Earth.

9. A force
   a. is expressed in newtons.
   b. can cause an object to speed up, slow down, or change direction.
   c. is a push or a pull.
   d. All of the above

CHAPTER 5

10. The amount of gravity between 1 kg of lead and Earth is _____ the amount of gravity between 1 kg of marshmallows and Earth.

    a. greater than   **c.** the same as

    b. less than   **d.** None of the above

Short Answer

11. Describe the relationship between motion and a reference point.

_____

_____

_____

_____

_____

_____

12. How is it possible to be accelerating and traveling at a constant speed?

_____

_____

_____

_____

_____

_____

13. Explain the difference between mass and weight.

_____

_____

_____

_____

_____

_____

_____

CONCEPT MAPPING

14. Use the following terms to create a concept map: *speed, velocity, acceleration, force, direction, motion.*

CRITICAL THINKING AND PROBLEM SOLVING

15. Your family is moving, and you are asked to help move some boxes. One box is so heavy that you must push it across the room rather than lift it. What are some ways you could reduce friction to make moving the box easier?

_____

_____

_____

16. Explain how using the term *accelerator* when talking about a car's gas pedal can lead to confusion, considering the scientific meaning of the word *acceleration.*

_____

_____

_____

CHAPTER 5

17. Explain why it is important for airplane pilots to know wind velocity, not just wind speed, during a flight.

_____

_____

_____

_____

_____

_____

_____

_____

MATH IN SCIENCE

18. A kangaroo hops 60 m to the east in 5 seconds.

   a. What is the kangaroo's speed?

_____

   b. What is the kangaroo's velocity?

_____

   c. The kangaroo stops at a lake for a drink of water, then starts hopping again to the south. Every second, the kangaroo's velocity increases 2.5 m/s. What is the kangaroo's acceleration after 5 seconds?

_____

INTERPRETING GRAPHICS

19. Is this a graph of positive or negative acceleration? How can you tell?

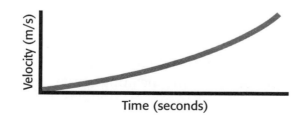

_____

Matter in Motion, continued

20. You know how to combine two forces that act in one or two directions. The same method you learned can be used to combine several forces acting in several directions. Examine the diagrams below, and predict with how much force and in what direction the object will move.

a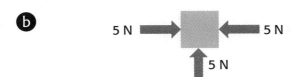

a. _____

b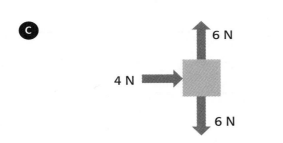

b. _____

c

c. _____

NOW WHAT DO YOU THINK?

Take a minute to review your answers to the ScienceLog questions at the beginning of the chapter. Have your answers changed? If necessary, revise your answers based on what you have learned since you began this chapter. Record your revisions in your ScienceLog.

CHAPTER

6 VOCABULARY & NOTES WORKSHEET

Forces in Motion

By studying the Vocabulary and Notes listed for each section below, you can gain a better understanding of this chapter.

SECTION 1

Vocabulary

In your own words, write a definition for each of the following terms in the space provided.

1. terminal velocity _____

_____

_____

_____

_____

_____

2. free fall _____

_____

_____

_____

_____

_____

3. projectile motion _____

_____

_____

_____

_____

_____

_____

Notes

Read the following section highlights. Then, in your own words, write the highlights in your ScienceLog.

- All objects accelerate toward Earth at 9.8 m/s/s.
- Air resistance slows the acceleration of falling objects.
- An object is in free fall if gravity is the only force acting on it.
- An orbit is formed by combining forward motion and free fall.
- Objects in orbit appear to be weightless because they are in free fall.
- A centripetal force is needed to keep objects in circular motion. Gravity acts as a centripetal force to keep objects in orbit.
- Projectile motion is the curved path an object follows when thrown or propelled near the surface of Earth.
- Projectile motion has two components—horizontal and vertical. Gravity affects only the vertical motion of projectile motion.

SECTION 2

Vocabulary

In your own words, write a definition for each of the following terms in the space provided.

1. inertia _____

_____

_____

2. momentum _____

_____

_____

Notes

Read the following section highlights. Then, in your own words, write the highlights in your ScienceLog.

- Newton's first law of motion states that the motion of an object will not change if no unbalanced forces act on it.
- Inertia is the tendency of matter to resist a change in motion. Mass is a measure of inertia.
- Newton's second law of motion states that the acceleration of an object depends on its mass and on the force exerted on it.
- Newton's third law of motion states that whenever one object exerts a force on a second object, the second object exerts an equal and opposite force on the first.
- Momentum is the property of a moving object that depends on its mass and velocity.
- When two or more objects interact, momentum may be exchanged, but the total amount of momentum does not change. This is the law of conservation of momentum.

Forces in Motion

USING VOCABULARY

To complete the following sentences, choose the correct term from each pair of terms listed below, and write the term in the space provided.

1. An object in motion tends to stay in motion because it has

_____ . (inertia or terminal velocity)

2. Falling objects stop accelerating at _____ .
(free fall or terminal velocity)

3. _____ is the path that a thrown object follows.
(Free fall or Projectile motion)

4. A property of moving objects that depends on mass and velocity is

_____ . (inertia or momentum)

5. _____ only occurs when there is no air resistance.
(Momentum or Free fall)

UNDERSTANDING CONCEPTS

Multiple Choice

6. A feather and a rock dropped at the same time from the same height would land at the same time when dropped by
  a. Galileo in Italy.
  b. Newton in England.
  c. an astronaut on the moon.
  d. an astronaut on the space shuttle.

7. When a soccer ball is kicked, the action and reaction forces do not cancel each other out because
  a. the force of the foot on the ball is bigger than the force of the ball on the foot.
  b. the forces act on two different objects.
  c. the forces act at different times.
  d. All of the above

8. An object is in projectile motion if
  a. it is thrown with a horizontal push.
  b. it is accelerated downward by gravity.
  c. it does not accelerate horizontally.
  d. All of the above

9. Newton's first law of motion applies
  a. to moving objects.
  b. to objects that are not moving.
  c. to objects that are accelerating.
  d. Both (a) and (b)

10. Acceleration of an object
   a. decreases as the mass of the object increases.
   b. increases as the force on the object increases.
   c. is in the same direction as the force on the object.
   d. All of the above

11. A golf ball and a bowling ball are moving at the same velocity. Which has more momentum?
   a. the golf ball, because it has less mass
   b. the bowling ball, because it has more mass
   c. They both have the same momentum because they have the same velocity.
   d. There is no way to know without additional information.

Short Answer

12. Explain how an orbit around the Earth is formed.

_____

_____

_____

_____

_____

13. Describe how gravity and air resistance combine when an object reaches terminal velocity.

_____

_____

_____

_____

14. Explain why friction can make observing Newton's first law of motion difficult.

_____

_____

_____

_____

CONCEPT MAPPING

15. Use the following terms to create a concept map: *gravity, free fall, terminal velocity, projectile motion, air resistance.*

CRITICAL THINKING AND PROBLEM SOLVING

16. During a shuttle launch, about 830,000 kg of fuel is burned in 8 minutes. The fuel provides the shuttle with a constant thrust, or push off the ground. How does Newton's second law of motion explain why the shuttle's acceleration increases during takeoff?

_____

_____

_____

_____

_____

_____

_____

17. When using a hammer to drive a nail into wood, you have to swing the hammer through the air with a certain velocity. Because the hammer has both mass and velocity, it has momentum. Describe what happens to the hammer's momentum after the hammer hits the nail.

_____

_____

_____

_____

_____

_____

_____

_____

_____

_____

Forces in Motion, continued

_____

18. Suppose you are standing on a skateboard or on in-line skates and you toss a back-pack full of heavy books toward your friend. What do you think will happen to you and why? Explain your answer in terms of Newton's third law of motion.

_____

_____

_____

_____

_____

_____

_____

_____

MATH IN SCIENCE

19. A 12 kg rock falls from rest off a cliff and hits the ground in 1.5 seconds.

    a. Ignoring air resistance, what is the rock's velocity just before it hits the ground?

_____

_____

_____

_____

_____

    b. What is the rock's weight after it hits the ground? (Hint: Weight is a measure of the gravitational force on an object.)

_____

_____

_____

_____

Forces in Motion, continued

INTERPRETING GRAPHICS

20. The picture below shows a common desk toy. If you pull one ball up and release it, it hits the balls at the bottom and comes to a stop. In the same instant, the ball on the other side swings up and repeats the cycle. How does conservation of momentum explain how this toy works?

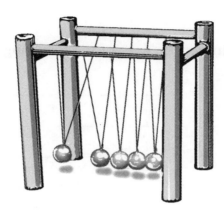

_____

_____

_____

_____

_____

_____

_____

_____

_____

_____

_____

NOW WHAT DO YOU THINK?

Take a minute to review your answers to the ScienceLog questions at the beginning of this chapter. Have your answers changed? If necessary, revise your answers based on what you have learned since you began this chapter. Record your revisions in your ScienceLog.

CHAPTER

7 VOCABULARY & NOTES WORKSHEET

Forces in Fluids

By studying the Vocabulary and Notes listed for each section below, you can gain a better understanding of this chapter.

SECTION 1

Vocabulary

In your own words, write a definition for each of the following terms in the space provided.

1. fluid _____

_____

_____

2. pressure _____

_____

_____

3. pascal _____

_____

_____

_____

4. atmospheric pressure _____

_____

5. density _____

_____

6. Pascal's principle _____

_____

_____

Notes

Read the following section highlights. Then, in your own words, write the highlights in your ScienceLog.

• A fluid is any material that flows and that takes the shape of its container.

• Pressure is force exerted on a given area.

- Moving particles of matter create pressure by colliding with one another and with the walls of their container.
- Fluids exert pressure equally in all directions.
- The pressure caused by the weight of Earth's atmosphere is called atmospheric pressure.
- Fluid pressure increases as depth increases.
- Fluids flow from areas of high pressure to areas of low pressure.
- Pascal's principle states that a change in pressure at any point in an enclosed fluid will be transmitted equally to all parts of the fluid.
- Hydraulic devices transmit changes of pressure through liquids.

SECTION 2

Vocabulary

In your own words, write a definition for each of the following terms in the space provided.

1. buoyant force _____

_____

_____

_____

_____

_____

2. Archimedes' principle _____

_____

_____

_____

_____

_____

Notes

Read the following section highlights. Then, in your own words, write the highlights in your ScienceLog.

- All fluids exert an upward force called buoyant force.
- Buoyant force is caused by differences in fluid pressure.
- Archimedes' principle states that the buoyant force on an object is equal to the weight of the fluid displaced by the object.
- Any object that is more dense than the surrounding fluid will sink; any object that is less dense than the surrounding fluid will float.

SECTION 3
Vocabulary

In your own words, write a definition for each of the following terms in the space provided.

1. Bernoulli's principle _____

_____

_____

_____

2. lift _____

_____

_____

_____

3. thrust _____

_____

_____

_____

4. drag _____

_____

_____

_____

Notes

Read the following section highlights. Then, in your own words, write the highlights in your ScienceLog.

- Bernoulli's principle states that fluid pressure decreases as the speed of a moving fluid increases.
- Wings are often shaped to allow airplanes to take advantage of decreased pressure in moving air in order to achieve flight.
- Lift is an upward force that acts against gravity.
- Lift on an airplane is determined by wing size and thrust (the forward force produced by the engine).
- Drag opposes motion through fluids.

CHAPTER

7 CHAPTER REVIEW WORKSHEET

Forces in Fluids

USING VOCABULARY

To complete the following sentences, choose the correct term from each of the pair of terms listed below, and write the term in the space provided.

1. _____ increases with the depth of a fluid.
(Pressure or Lift)

2. A plane's engine produces _____ to push the plane forward.
(thrust or drag)

3. Force divided by area is known as _____ . (density or pressure)

4. The hydraulic brakes of a car transmit pressure through fluid. This is an example of

_____ . (Archimedes' principle or Pascal's principle)

5. Bernoulli's principle states that the pressure exerted by a moving fluid is

_____ the pressure of the fluid when it is not moving.
(greater than or less than)

UNDERSTANDING CONCEPTS

Multiple Choice

6. The curve on the top of a wing
   a. causes air to travel farther in the same amount of time as the air below the wing.
   b. helps create lift.
   c. creates a low-pressure zone above the wing.
   d. All of the above

7. An object displaces a volume of fluid that
   a. is equal to its own volume.
   b. is less than its own volume.
   c. is greater than its own volume.
   d. is more dense than itself.

8. Fluid pressure is always directed
   a. up.   **c.** sideways.
   b. down.   **d.** in all directions.

9. If an object weighing 50 N displaces a volume of water with a weight of 10 N, what is the buoyant force on the object?
   a. 60 N
   b. 50 N
   c. 40 N
   d. 10 N

_____

Forces in Fluids, continued

_____

10. A helium-filled balloon will float in air because

    a. there is more air than helium.

    b. helium is less dense than air.

    c. helium is as dense as air.

    d. helium is more dense than air.

11. Materials that can flow to fit their containers include

    a. gases.

    b. liquids.

    c. both gases and liquids.

    d. neither gases nor liquids.

Short Answer

12. What two factors determine the amount of lift achieved by an airplane?

_____

_____

13. Where is water pressure greater, at a depth of 1 m in a large lake or at a depth of 2 m in a small pond? Explain.

_____

_____

_____

_____

14. Is there buoyant force on an object at the bottom of an ocean? Explain your reasoning.

_____

_____

_____

_____

15. Why are liquids used in hydraulic brakes instead of gases?

_____

_____

_____

_____

Forces in Fluids, continued

CONCEPT MAPPING

16. Use the following terms to create a concept map: *fluid, pressure, depth, buoyant force, density.*

CRITICAL THINKING AND PROBLEM SOLVING

17. Compared with an empty ship, will a ship loaded with plastic-foam balls float higher or lower in the water? Explain your reasoning.

_____

_____

_____

_____

_____

_____

18. Inside all vacuum cleaners is a high-speed fan. Explain how this fan causes dirt to be picked up by the vacuum cleaner.

_____

_____

_____

_____

19. A 600 N clown on stilts says to two 600 N clowns sitting on the ground, "I am exerting twice as much pressure as the two of you together!" Could this statement be true? Explain your reasoning.

_____

_____

_____

_____

_____

_____

_____

_____

MATH IN SCIENCE

20. Calculate the area of a 1,500 N object that exerts a pressure of 500 Pa (N/m²). Then calculate the pressure exerted by the same object over twice that area. Be sure to express your answers in the correct SI unit.

_____

_____

_____

_____

Forces in Fluids, continued

INTERPRETING GRAPHICS

Examine the illustration of an iceberg below, and answer the questions that follow.

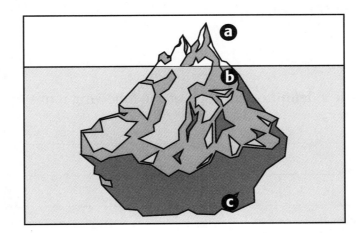

21. At what point (a, b, or c) is water pressure greatest on the iceberg? _____

22. How much of the iceberg has a weight equal to the buoyant force?
  a. all of it
  b. the section from a to b
  c. the section from b to c

23. How does the density of ice compare with the density of water?

_____

24. Why do you think icebergs are so dangerous to passing ships?

_____

_____

_____

_____

_____

_____

NOW WHAT DO YOU THINK?

Take a minute to review your answers to the ScienceLog questions at the beginning of this chapter. Have your answers changed? If necessary, revise your answers based on what you have learned since you began this chapter. Record your revisions in your ScienceLog.

Work and Machines

By studying the Vocabulary and Notes listed for each section below, you can gain a better understanding of this chapter.

SECTION 1

Vocabulary

In your own words, write a definition for each of the following terms in the space provided.

1. work _____

_____

_____

2. joule _____

_____

_____

3. power _____

_____

_____

4. watt _____

_____

_____

Notes

Read the following section highlights. Then, in your own words, write the highlights in your ScienceLog.

• Work occurs when a force causes an object to move in the direction of the force. The unit for work is the joule (J).

• Work is done on an object only when a force makes an object move and only while that force is applied.

• For work to be done on an object, the direction of the object's motion must be in the same direction as the force applied.

• Work can be calculated by multiplying force by distance.

• Power is the rate at which work is done. The unit for power is the watt (W).

• Power can be calculated by dividing the amount of work by the time taken to do that work.

Work and Machines, continued

SECTION 2

Vocabulary

In your own words, write a definition for each of the following terms in the space provided.

1. machine _____

_____

_____

2. work input _____

_____

_____

3. work output _____

_____

_____

4. mechanical advantage _____

_____

_____

5. mechanical efficiency _____

_____

_____

_____

Notes

Read the following section highlights. Then, in your own words, write the highlights in your ScienceLog.

• A machine makes work easier by changing the size or direction (or both) of a force.

• When a machine changes the size of a force, the distance through which the force is exerted must also change. Force or distance can increase, but not together.

• Mechanical advantage tells how many times a machine multiplies force. It can be calculated by dividing the output force by the input force.

• Mechanical efficiency is a comparison of a machine's work output with work input. Mechanical efficiency is calculated by dividing work output by work input and is expressed as a percentage.

• Machines are not 100 percent efficient because some of the work done by a machine is used to overcome friction. So work output is always less than work input.

CHAPTER 8

SECTION 3
Vocabulary

In your own words, write a definition for each of the following terms in the space provided.

1. lever _____

_____

_____

_____

_____

_____

_____

2. inclined plane _____

_____

_____

_____

_____

3. wedge _____

_____

_____

4. screw _____

_____

_____

_____

_____

Work and Machines, continued

5. wheel and axle _____

_____

_____

_____

_____

_____

_____

6. pulley _____

_____

_____

_____

_____

_____

_____

7. compound machine _____

_____

_____

_____

_____

_____

Notes

Read the following section highlights. Then, in your own words, write the highlights in your ScienceLog.

- All machines are constructed from these six simple machines: lever, inclined plane, wedge, screw, wheel and axle, and pulley.
- Compound machines consist of two or more simple machines.
- Compound machines have low mechanical efficiencies because they have more moving parts and thus more friction to overcome.

CHAPTER 8

CHAPTER

8 **CHAPTER REVIEW WORKSHEET**

Work and Machines

USING VOCABULARY

For each pair of terms, explain the difference in their meanings.

1. joule/watt _____

_____

_____

2. work output/work input _____

_____

_____

_____

3. mechanical efficiency/mechanical advantage _____

_____

_____

_____

_____

_____

_____

4. screw/inclined plane _____

_____

_____

_____

5. simple machine/compound machine _____

_____

_____

_____

_____

Work and Machines, continued

UNDERSTANDING CONCEPTS

Multiple Choice

6. Work is being done when

    a. you apply a force to an object.

    b. an object is moving after you apply a force to it.

    c. you exert a force that moves an object in the direction of the force.

    d. you do something that is difficult.

7. The work output for a machine is always less than the work input because

    a. all machines have a mechanical advantage.

    b. some of the work done is used to overcome friction.

    c. some of the work done is used to overcome distance.

    d. power is the rate at which work is done.

8. The unit for work is the

    a. joule.

    b. joule per second.

    c. newton.

    d. watt.

9. Which of the following is not a simple machine?

    a. a faucet handle

    b. a jar lid

    c. a can opener

    d. a seesaw

10. Power is

    a. how strong someone or something is.

    b. how much force is being used.

    c. how much work is being done.

    d. how fast work is being done.

11. The unit for power is the

    a. newton.

    b. kilogram.

    c. watt.

    d. joule.

12. A machine can increase

    a. distance at the expense of force.

    b. force at the expense of distance.

    c. neither distance nor force.

    d. Both (a) and (b)

Short Answer

13. Identify the simple machines that make up a pair of scissors.

_____

_____

_____

CHAPTER 8

_____
Work and Machines, continued
_____

14. In two or three sentences, explain the force-distance trade-off that occurs when a machine is used to make work easier.

_____

_____

_____

_____

_____

_____

15. Explain why you do work on a bag of groceries when you pick it up but not when you carry it.

_____

_____

_____

_____

_____

CONCEPT MAPPING

16. Create a concept map using the following terms: *work, force, distance, machine, mechanical advantage.*

Work and Machines, continued

CRITICAL THINKING AND PROBLEM SOLVING

17. Why do you think levers usually have a greater mechanical efficiency than other simple machines do?

_____

_____

_____

_____

_____

_____

_____

18. A winding road is actually a series of inclined planes. Describe how a winding road makes it easier for vehicles to travel up a hill.

_____

_____

_____

_____

_____

_____

_____

19. Why do you think you would not want to reduce the friction involved in using a winding road?

_____

_____

_____

_____

_____

_____

CHAPTER 8

MATH IN SCIENCE

20. You and a friend together apply a force of 1,000 N to a 3,000 N automobile to make it roll 10 m in 1 minute and 40 seconds.

    a. How much work did you and your friend do together?

_____

    b. What was your combined power?

_____

INTERPRETING GRAPHICS

For each of the images below, identify the class of lever used and calculate the mechanical advantage.

21.

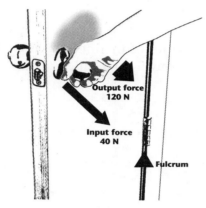

_____

22.

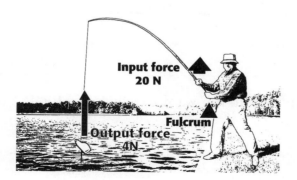

NOW WHAT DO YOU THINK?

Take a minute to review your answers to the ScienceLog questions at the beginning of this chapter. Have your answers changed? If necessary, revise your answers based on what you have learned since you began this chapter. Record your revisions in your ScienceLog.

CHAPTER

9 VOCABULARY & NOTES WORKSHEET

Energy and Energy Resources

By studying the Vocabulary and Notes listed for each section below, you can gain a better understanding of this chapter.

SECTION 1

Vocabulary

In your own words, write a definition for each of the following terms in the space provided.

1. energy _____

_____

_____

2. kinetic energy _____

_____

_____

3. potential energy _____

_____

_____

4. mechanical energy _____

_____

_____

Notes

Read the following section highlights. Then, in your own words, write the highlights in your ScienceLog.

- Energy is the ability to do work, and work is the transfer of energy. Both energy and work are expressed in joules.
- Kinetic energy is energy of motion and depends on speed and mass.
- Potential energy is energy of position or shape. Gravitational potential energy depends on weight and height.
- Mechanical energy is the sum of kinetic energy and potential energy.
- Thermal energy, sound energy, electrical energy, and light energy can all be forms of kinetic energy.
- Chemical energy, electrical energy, sound energy, and nuclear energy can all be forms of potential energy.

▲ **CHAPTER 9**

SECTION 2

Vocabulary

In your own words, write a definition of the following term in the space provided.

1. energy conversion _____

_____

_____

_____

_____

Notes

Read the following section highlights. Then, in your own words, write the highlights in your ScienceLog.

• An energy conversion is a change from one form of energy to another. Any form of energy can be converted into any other form of energy.

• Machines can transfer energy and convert energy into a more useful form.

• Energy conversions help to make energy useful by changing energy into the form you need.

SECTION 3

Vocabulary

In your own words, write a definition for each of the following terms in the space provided.

1. friction _____

_____

_____

_____

_____

2. law of conservation of energy _____

_____

_____

_____

_____

Notes

Read the following section highlights. Then, in your own words, write the highlights in your ScienceLog.

- Because of friction, some energy is always converted into thermal energy during an energy conversion.
- Energy is conserved within a closed system. According to the law of conservation of energy, energy can be neither created nor destroyed.
- Perpetual motion is impossible because some of the energy put into a machine will eventually be converted completely into thermal energy due to friction.

SECTION 4

Vocabulary

In your own words, write a definition for each of the following terms in the space provided.

1. energy resource _____

_____

_____

_____

_____

_____

_____

2. nonrenewable resources _____

_____

_____

_____

_____

_____

_____

_____

CHAPTER 9

3. fossil fuels _____

_____

_____

_____

_____

_____

_____

4. renewable resources _____

_____

_____

_____

_____

_____

_____

_____

_____

Notes

Read the following section highlights. Then, in your own words, write the highlights in your ScienceLog.

• An energy resource is a natural resource that can be converted into other forms of energy in order to do useful work.

• Nonrenewable resources cannot be replaced after they are used or can only be replaced after long periods of time. They include fossil fuels and nuclear energy.

• Fossil fuels are nonrenewable resources formed from the remains of ancient organisms. Coal, petroleum, and natural gas are fossil fuels.

• Renewable resources can be used and replaced in nature over a relatively short period of time. They include solar energy, wind energy, energy from water, geothermal energy, and biomass.

• The sun is the source of most energy on Earth.

• Depending on where you live and what you need energy for, one energy resource can be a better choice than another.

CHAPTER

Energy and Energy Resources

USING VOCABULARY

For each pair of terms, explain the difference in their meanings.

1. potential energy/kinetic energy _____

_____

_____

_____

2. friction/energy conversion _____

_____

_____

_____

_____

_____

3. energy conversion/law of conservation of energy _____

_____

_____

_____

_____

_____

_____

4. energy resources/fossil fuels _____

_____

_____

_____

_____

_____

5. renewable resources/nonrenewable resources _____

_____

_____

_____

_____

UNDERSTANDING CONCEPTS

Multiple Choice

6. Kinetic energy depends on
   a. mass and volume.
   b. speed and weight.
   c. weight and height.
   d. speed and mass.

7. Gravitational potential energy depends on
   a. mass and speed.
   b. weight and height.
   c. mass and weight.
   d. height and distance.

8. Which of the following is not a renewable resource?
   a. wind energy
   b. nuclear energy
   c. solar energy
   d. geothermal energy

9. Which of the following is a conversion from chemical energy to thermal energy?
   a. Food is digested and used to regulate body temperature.
   b. Charcoal is burned in a barbecue pit.
   c. Coal is burned to boil water.
   d. All of the above

10. Machines can
   a. increase energy.
   b. transfer energy.
   c. convert energy.
   d. Both (b) and (c)

11. In every energy conversion, some energy is always converted into
   a. kinetic energy.
   b. potential energy.
   c. thermal energy.
   d. mechanical energy.

12. An object that has kinetic energy must be
   a. at rest.
   b. lifted above the Earth's surface.
   c. in motion.
   d. None of the above

13. Which of the following is not a fossil fuel?
   a. gasoline   **c.** firewood
   b. coal   **d.** natural gas

Short Answer

14. Name two forms of energy, and relate them to kinetic or potential energy.

_____

_____

_____

_____

_____

_____

_____

_____

15. Give three specific examples of energy conversions.

_____

_____

_____

_____

_____

_____

_____

16. Explain how energy is conserved within a closed system.

_____

_____

_____

Energy and Energy Resources, continued

17. How are fossil fuels formed?

_____

_____

_____

_____

_____

_____

CONCEPT MAPPING

18. Use the following terms to create a concept map: *energy, machines, energy conversions, thermal energy, friction.*

Energy and Energy Resources, continued

CRITICAL THINKING AND PROBLEM SOLVING

19. What happens when you blow up a balloon and release it? Describe what you would see in terms of energy.

_____

_____

_____

_____

20. After you coast down a hill on your bike, you eventually come to a complete stop unless you keep pedaling. Relate this to the reason why perpetual motion is impossible.

_____

_____

_____

21. Look at the picture of the pole-vaulter below. Trace the energy conversions involved in this event, beginning with the pole-vaulter's breakfast of an orange-banana smoothie.

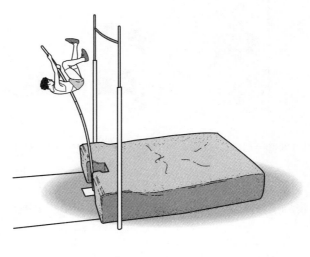

_____

_____

_____

_____

Energy and Energy Resources, continued

---

22. If the sun were exhausted of its nuclear energy, what would happen to our energy resources on Earth?

_____

_____

MATH IN SCIENCE

23. A box has 400 J of gravitational potential energy.

    a. How much work had to be done to give the box that energy?

_____

    b. If the box weighs 100 N, how far was it lifted?

_____

INTERPRETING GRAPHICS

24. Look at the illustration below, and answer the questions that follow.

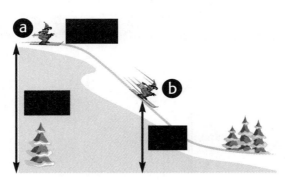

    a. What is the skier's gravitational potential energy at point *A*?

_____

    b. What is the skier's gravitational potential energy at point *B*?

_____

    c. What is the skier's kinetic energy at point *B*? (Hint: mechanical energy = potential energy + kinetic energy.)

_____

NOW WHAT DO YOU THINK?

Take a minute to review your answers to the ScienceLog questions at the beginning of the chapter. Have your answers changed? If necessary, revise your answers based on what you have learned since you began this chapter. Record your revisions in your ScienceLog.

CHAPTER

10 VOCABULARY & NOTES WORKSHEET

Heat and Heat Technology

By studying the Vocabulary and Notes listed for each section below, you can gain a better understanding of this chapter.

SECTION 1

Vocabulary

In your own words, write a definition of each of the following terms in the space provided.

1. temperature _____

_____

_____

_____

2. thermal expansion _____

_____

_____

_____

3. absolute zero _____

_____

_____

_____

Notes

Read the following section highlights. Then, in your own words, write the highlights in your ScienceLog.

- Temperature is a measure of the average kinetic energy of the particles of a substance. It is a specific measurement of how hot or cold a substance is.
- Thermal expansion is the increase in volume of a substance due to an increase in temperature. Temperature is measured according to the expansion of the liquid in a thermometer.
- Fahrenheit, Celsius, and Kelvin are three temperature scales.
- Absolute zero—0 K, or –273°C—is the lowest possible temperature.
- A thermostat works according to the thermal expansion of a bimetallic strip.

CHAPTER 10

SECTION 2

Vocabulary

In your own words, write a definition of each of the following terms in the space provided.

1. heat _____

_____

_____

_____

2. thermal energy _____

_____

_____

_____

3. conduction _____

_____

_____

_____

4. conductor _____

_____

_____

_____

5. insulator _____

_____

_____

_____

6. convection _____

_____

_____

_____

7. radiation _____

_____

_____

8. specific heat capacity _____

_____

_____

_____

Notes

Read the following section highlights. Then, in your own words, write the highlights in your ScienceLog.

• Heat is the transfer of energy between objects that are at different temperatures.

• Thermal energy is the total kinetic energy of the particles that make up a substance.

• Energy transfer will always occur from higher temperatures to lower temperatures until thermal equilibrium is reached.

• Conduction, convection, and radiation are three methods of energy transfer.

• Specific heat capacity is the amount of energy needed to change the temperature of 1 kg of a substance by 1°C. Different substances have different specific heat capacities.

• Energy transferred by heat cannot be measured directly. It must be calculated using specific heat capacity, mass, and change in temperature.

• A calorimeter is used to determine the specific heat capacity of a substance.

SECTION 3

Vocabulary

In your own words, write a definition of each of the following terms in the space provided.

1. states of matter _____

_____

_____

_____

2. change of state _____

_____

_____

_____

CHAPTER 10

Notes

Read the following section highlights. Then, in your own words, write the highlights in your ScienceLog.

- A substance's state is determined by the speed of its particles and the attraction between them.
- Thermal energy transferred during a change of state does not change a substance's temperature. Rather, it causes a substance's particles to be rearranged.
- Chemical changes can cause thermal energy to be absorbed or released.

SECTION 4

Vocabulary

In your own words, write a definition of each of the following terms in the space provided.

1. insulation _____

_____

_____

2. heat engine _____

_____

_____

3. thermal pollution _____

_____

_____

Notes

Read the following section highlights. Then, in your own words, write the highlights in your ScienceLog.

- Central heating systems include hot-water heating systems and warm-air heating systems.
- Solar heating systems can be passive or active.
- Heat engines use heat to do work. External combustion engines burn fuel outside the engine. Internal combustion engines burn fuel inside the engine.
- A cooling system transfers thermal energy from cooler temperatures to warmer temperatures by doing work.
- Transferring excess thermal energy to lakes and rivers can result in thermal pollution.

CHAPTER

10 CHAPTER REVIEW WORKSHEET

Heat and Heat Technology

USING VOCABULARY

For each pair of terms, explain the difference in their meanings.

1. temperature/thermal energy _____

_____

_____

_____

_____

_____

2. heat/thermal energy _____

_____

_____

_____

_____

3. conductor/insulator _____

_____

_____

_____

4. conduction/convection _____

_____

_____

_____

5. states of matter/change of state _____

_____

_____

_____

CHAPTER 10

UNDERSTANDING CONCEPTS

Multiple Choice

6. Which of the following temperatures is the lowest?

    a. 100°C

    b. 100°F

    c. 100 K

    d. They are the same.

7. Compared with the Pacific Ocean, a cup of hot chocolate has

    a. more thermal energy and a higher temperature.

    b. less thermal energy and a higher temperature.

    c. more thermal energy and a lower temperature.

    d. less thermal energy and a lower temperature.

8. The energy units on a food label are

    a. degrees

    b. Calories

    c. calories

    d. joules

9. Which of the following materials would not be a good insulator?

    a. wood

    b. cloth

    c. metal

    d. rubber

10. The engine in a car is a(n)

    a. heat engine.

    b. external combustion engine.

    c. internal combustion engine.

    d. Both (a) and (c)

11. Materials that warm up or cool down very quickly have a

    a. low specific heat capacity.

    b. high specific heat capacity.

    c. low temperature.

    d. high temperature.

12. In an air conditioner, thermal energy is

    a. transferred from higher to lower temperatures.

    b. transferred from lower to higher temperatures.

    c. used to do work.

    d. taken from air outside a building and transferred to air inside the building.

Short Answer

13. How does temperature relate to kinetic energy?

_____

_____

_____

_____

Heat and Heat Technology, continued

14. What is specific heat capacity?

_____

_____

_____

_____

_____

_____

_____

15. Explain how heat affects matter during a change of state.

_____

_____

_____

_____

_____

_____

_____

_____

16. Describe how a bimetallic strip works in a thermostat.

_____

_____

_____

_____

_____

_____

_____

CHAPTER 10

CONCEPT MAPPING

17. Use the following terms to create a concept map: *thermal energy, temperature, radiation, heat, conduction, convection.*

CRITICAL THINKING AND PROBLEM SOLVING

Write one or two sentences to answer the following questions:

18. Why does placing a jar under warm running water help to loosen the lid on the jar?

_____

_____

_____

Heat and Heat Technology, continued

19. Why do you think a down-filled jacket keeps you so warm? (Hint: Think about what insulation does.)

_____

_____

_____

_____

20. Would opening the refrigerator cool a room in a house? Why or why not?

_____

_____

_____

_____

_____

21. In a hot-air balloon, air is heated by a flame. Explain how this enables the balloon to float in the air.

_____

_____

_____

MATH IN SCIENCE

22. The weather forecast calls for a temperature of 86°F. What is the corresponding temperature in degrees Celsius? in kelvins?

_____

_____

23. Suppose 1,300 mL of water are heated from 20°C to 100°C. How much energy was transferred to the water? (Hint: Water's specific heat capacity is 4,184 J/kg •°C.)

_____

_____

CHAPTER 10

INTERPRETING GRAPHICS

Examine the graph below, and then answer the questions that follow.

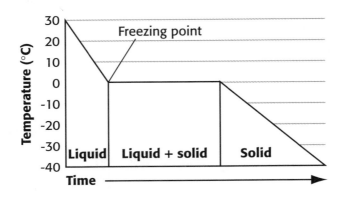

24. What physical change does this graph illustrate?

_____

_____

_____

25. What is the freezing point of this liquid?

_____

_____

26. What is happening at the point where the line is horizontal?

_____

_____

_____

_____

NOW WHAT DO YOU THINK?

Take a minute to review your answers to the ScienceLog questions at the beginning of this chapter. Have your answers changed? If necessary, revise your answers based on what you have learned since you began this chapter.

CHAPTER

11 VOCABULARY & NOTES WORKSHEET

Introduction to Atoms

By studying the Vocabulary and Notes listed for each section below, you can gain a better understanding of this chapter.

SECTION 1

Vocabulary

In your own words, write a definition for each of the following terms in the space provided.

1. atom _____

_____

2. theory _____

_____

_____

3. electrons _____

_____

4. model _____

_____

5. nucleus _____

_____

6. electron clouds _____

_____

Notes

Read the following section highlights. Then, in your own words, write the highlights in your ScienceLog.

• Atoms are the smallest particles of an element that retain the properties of the element.

• In ancient Greece, Democritus argued that atoms were the smallest particles in all matter.

• Dalton proposed an atomic theory that stated the following: Atoms are small particles that make up all matter; atoms cannot be created, divided, or destroyed; atoms of an element are exactly alike; atoms of different elements are different; and atoms join together to make new substances.

• Thomson discovered electrons. His plum-pudding model described the atom as a lump of positively charged material with negative electrons scattered throughout.

- Rutherford discovered that atoms contain a small, dense, positively charged center called the nucleus.
- Bohr suggested that electrons move around the nucleus at only certain distances.
- According to the current atomic theory, electron clouds are where electrons are most likely to be in the space around the nucleus.

SECTION 2

Vocabulary

In your own words, write a definition for each of the following terms in the space provided.

1. protons _____

_____

_____

_____

_____

2. atomic mass unit (amu) _____

_____

_____

_____

_____

3. neutrons _____

_____

_____

_____

_____

4. atomic number _____

_____

_____

_____

_____

5. isotopes _____

_____

_____

_____

_____

6. mass number _____

_____

_____

_____

_____

7. atomic mass _____

_____

_____

_____

_____

_____

Notes

Read the following section highlights. Then, in your own words, write the highlights in your ScienceLog.

- A proton is a positively charged particle with a mass of 1 amu.
- A neutron is a particle with no charge that has a mass of 1 amu.
- An electron is a negatively charged particle with an extremely small mass.
- Protons and neutrons make up the nucleus. Electrons are found in electron clouds outside the nucleus.
- The number of protons in the nucleus of an atom is the atomic number. The atomic number identifies the atoms of a particular element.
- Isotopes of an atom have the same number of protons but have different numbers of neutrons. Isotopes share most of the same chemical and physical properties.
- The mass number of an atom is the sum of the atom's neutrons and protons.
- The atomic mass is an average of the masses of all naturally occurring isotopes of an element.
- The four forces at work in an atom are gravity, the electromagnetic force, the strong force, and the weak force.

CHAPTER

11 CHAPTER REVIEW WORKSHEET

Introduction to Atoms

USING VOCABULARY

The statements below are false. For each statement, replace the underlined word to make a true statement.

1. Electrons are found in the <u>nucleus</u> of an atom.

_____

2. All atoms of the same element contain the same number of <u>neutrons</u>.

_____

3. <u>Protons</u> have no electrical charge.

_____

4. The <u>atomic number</u> of an element is the number of protons and neutrons in the nucleus.

_____

5. The <u>mass number</u> is an average of the masses of all naturally occurring isotopes of an element.

_____

UNDERSTANDING CONCEPTS

Multiple Choice

6. The discovery of which particle proved that the atom is not indivisible?

  a. proton   **c.** electron
  b. neutron   **d.** nucleus

7. In his gold foil experiment, Rutherford concluded that the atom is mostly empty space with a small, massive, positively charged center because

  a. most of the particles passed straight through the foil.
  b. some particles were slightly deflected.
  c. a few particles bounced back.
  d. All of the above

8. How many protons does an atom with an atomic number of 23 and a mass number of 51 have?

  a. 23   **c.** 51
  b. 28   **d.** 74

9. An atom has no overall charge if it contains equal numbers of

  a. electrons and protons.
  b. neutrons and protons.
  c. neutrons and electrons.
  d. None of the above

Introduction to Atoms, continued

10. Which statement about protons is true?

    a. Protons have a mass of 1/1840 amu.

    b. Protons have no charge.

    c. Protons are part of the nucleus of an atom.

    d. Protons circle the nucleus of an atom.

11. Which statement about neutrons is true?

    a. Neutrons have a mass of 1 amu.

    b. Neutrons circle the nucleus of an atom.

    c. Neutrons are the only particles that make up the nucleus.

    d. Neutrons have a negative charge.

12. Which of the following determines the identity of an element?

    a. atomic number   **c.** atomic mass

    b. mass number   **d.** overall charge

13. Isotopes exist because atoms of the same element can have different numbers of

    a. protons.

    b. neutrons.

    c. electrons.

    d. None of the above

Short Answer

14. Why do scientific theories change?

_____

_____

_____

_____

15. What force holds electrons in atoms?

_____

_____

16. In two or three sentences, describe the plum-pudding model of the atom.

_____

_____

_____

_____

CONCEPT MAPPING

17. Use the following terms to create a concept map: *atom, nucleus, protons, neutrons, electrons, isotopes, atomic number, mass number.*

CRITICAL THINKING AND PROBLEM SOLVING

18. Particle accelerators are devices that speed up charged particles in order to smash them together. Sometimes the result of the collision is a new nucleus. How can scientists determine whether the nucleus formed is that of a new element or that of a new isotope of a known element?

_____

_____

_____

_____

_____

_____

Introduction to Atoms, continued

_____

19. John Dalton made a number of statements about atoms that are now known to be incorrect. Why do you think his atomic theory is still found in science textbooks?

_____

_____

_____

_____

MATH IN SCIENCE

20. Calculate the atomic mass of gallium consisting of 60 percent gallium-69 and 40 percent gallium-71. Show your work.

_____

21. Calculate the number of protons, neutrons, and electrons in an atom of zirconium-90, which has an atomic number of 40.

_____

INTERPRETING GRAPHICS

22. Study the models below, and answer the questions that follow:

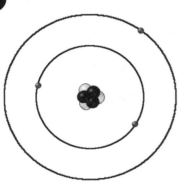

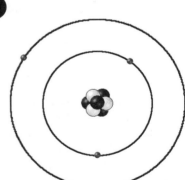

Key

○ Proton

● Neutron

· Electron

Introduction to Atoms, continued

---

a. Which models represent isotopes of the same element?

_____

b. What is the atomic number for (a)?

_____

c. What is the mass number for (b)?

_____

23. Predict how the direction of the moving particle in the figure below will change, and explain what causes the change to occur.

_____

_____

_____

_____

_____

_____

_____

NOW WHAT DO YOU THINK?

Take a minute to review your answers to the ScienceLog questions at the beginning of this chapter. Have your answers changed? If necessary, revise your answers based on what you have learned since you began this chapter. Record your revisions in your ScienceLog.

12 VOCABULARY & NOTES WORKSHEET

The Periodic Table

By studying the Vocabulary and Notes listed for each section below, you can gain a better understanding of this chapter.

SECTION 1

Vocabulary

In your own words, write a definition for each of the following terms in the space provided.

1. periodic _____

_____

_____

2. periodic law _____

_____

_____

_____

3. period _____

_____

_____

_____

4. group _____

_____

Notes

Read the following section highlights. Then in your own words, write the highlights in your ScienceLog.

• Mendeleev developed the first periodic table. He arranged elements in order of increasing atomic mass. The properties of elements repeated in an orderly pattern, allowing Mendeleev to predict properties for elements that had not yet been discovered.

• Moseley rearranged the elements in order of increasing atomic number.

• The periodic law states that the chemical and physical properties of elements are periodic functions of their atomic numbers.

• Elements in the periodic table are divided into metals, metalloids, and nonmetals.

• Each element has a chemical symbol that is recognized around the world.

• A horizontal row of elements is called a period. The elements gradually change from metallic to nonmetallic from left to right across each period.

• A vertical column of elements is called a group or family. Elements in a group usually have similar properties.

The Periodic Table, continued

SECTION 2

Vocabulary

In your own words, write a definition for each of the following terms in the space provided.

1. alkali metals _____

_____

_____

2. alkaline-earth metals _____

_____

_____

3. halogens _____

_____

_____

4. noble gases _____

_____

_____

Notes

Read the following section highlights. Then in your own words, write the highlights in your ScienceLog.

- The alkali metals (Group 1) are the most reactive metals. Atoms of the alkali metals have one electron in their outer level.
- The alkaline-earth metals (Group 2) are less reactive than the alkali metals. Atoms of the alkaline-earth metals have two electrons in their outer level.
- The transition metals (Groups 3–12) include most of the well-known metals as well as the lanthanides and actinides located below the periodic table.
- Groups 13–16 contain the metalloids along with some metals and nonmetals. The atoms of the elements in each of these groups have the same number of electrons in their outer level.
- The halogens (Group 17) are very reactive nonmetals. Atoms of the halogens have seven electrons in their outer level.
- The noble gases (Group 18) are unreactive nonmetals. Atoms of the noble gases have a complete set of electrons in their outer level.
- Hydrogen is set off by itself because its properties do not match the properties of any one group.

CHAPTER

12 CHAPTER REVIEW WORKSHEET

The Periodic Table

USING VOCABULARY

To complete the following sentences, choose the correct term from each pair of terms listed below, and write the term in the space provided.

1. Elements in the same vertical column in the periodic table belong to the same

_____ . (group or period)

2. Elements in the same horizontal row in the periodic table belong to the same

_____ . (group or period)

3. The most reactive metals are _____ .
(alkali metals or alkaline-earth metals)

4. Elements that are unreactive are called _____ .
(noble gases or halogens)

UNDERSTANDING CONCEPTS

Multiple Choice

5. An element that is a very reactive gas is most likely a member of the

    a. noble gases.   **c.** halogens.
    b. alkali metals.   **d.** actinides.

6. Which statement is true?

    a. Alkali metals are generally found in their uncombined form.
    b. Alkali metals are Group 1 elements.
    c. Alkali metals should be stored under water.
    d. Alkali metals are unreactive.

7. Which statement about the periodic table is false?

    a. There are more metals than nonmetals.
    b. The metalloids are located in Groups 13 through 16.
    c. The elements at the far left of the table are nonmetals.
    d. Elements are arranged by increasing atomic number.

8. One property of most nonmetals is that they are

    a. shiny.
    b. poor conductors of electric current.
    c. flattened when hit with a hammer.
    d. solids at room temperature.

9. Which is a true statement about elements?

    a. Every element occurs naturally.
    b. All elements are found in their uncombined form in nature.
    c. Each element has a unique atomic number.
    d. All of the elements exist in approximately equal quantities.

_____

The Periodic Table, continued

_____

10. Which is NOT found on the periodic table?
  a. The atomic number of each element
  b. The symbol of each element
  c. The density of each element
  d. The atomic mass of each element

Short Answer

11. Why was Mendeleev's periodic table useful?

_____

_____

_____

12. How is Moseley's basis for arranging the elements different from Mendeleev's?

_____

_____

_____

13. How is the periodic table like a calendar?

_____

_____

_____

14. Describe the location of metals, metalloids, and nonmetals on the periodic table.

_____

_____

_____

The Periodic Table, *continued*

CONCEPT MAPPING

15. Use the following terms to create a concept map: *periodic table, elements, groups, periods, metals, nonmetals, metalloids.*

CRITICAL THINKING AND PROBLEM SOLVING

16. When an element with 115 protons in its nucleus is synthesized, will it be a metal, a nonmetal, or a metalloid? Explain.

_____

17. Look at Mendeleev's periodic table in Figure 2, on page 303 of your textbook. Why was Mendeleev not able to make any predictions about the noble gas elements?

_____

_____

_____

_____

18. Your classmate offers to give you a piece of sodium he found while hiking. What is your response? Explain.

_____

_____

_____

The Periodic Table, continued

_____

19. Determine the identity of each element described below:

   a. This metal is very reactive, has properties similar to magnesium, and is in the same period as bromine.

   _____

   b. This nonmetal is in the same group as lead.

   _____

   c. This metal is the most reactive metal in its period and cannot be found uncombined in nature. Each atom of the element contains 19 protons.

   _____

MATH IN SCIENCE

20. The chart below shows the percentages of elements in the Earth's crust.

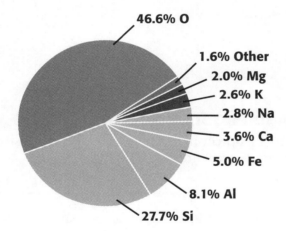

Excluding the "Other" category, what percentage of the Earth's crust is
a. alkali metals?

_____

b. alkaline-earth metals?

_____

The Periodic Table, continued

INTERPRETING GRAPHICS

21. Study the diagram below to determine the pattern of the images. Predict the missing image, and draw it. Identify which properties are periodic and which properties are shared within a group.

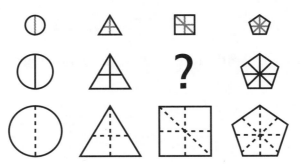

_____

_____

_____

_____

NOW WHAT DO YOU THINK?

Take a minute to review your answers to the ScienceLog questions at the beginning of this chapter. Have your answers changed? If necessary, revise your answers based on what you have learned since you began this chapter. Record your revisions in your ScienceLog.

CHAPTER 12

Chemical Bonding

By studying the Vocabulary and Notes listed for each section below, you can gain a better understanding of this chapter.

SECTION 1

Vocabulary

In your own words, write a definition for each of the following terms in the space provided.

1. chemical bonding _____

_____

_____

_____

2. chemical bond _____

_____

_____

_____

3. valence electrons _____

_____

_____

_____

Notes

Read the following section highlights. Then in your own words, write the highlights in your ScienceLog.

• Chemical bonding is the joining of atoms to form new substances. A chemical bond is a force of attraction that holds two atoms together.

• Valence electrons are the electrons in the outermost energy level of an atom. These electrons are used to form bonds.

• Most atoms form bonds by gaining, losing, or sharing electrons until they have 8 valence electrons. Atoms of hydrogen, lithium, and helium need only 2 electrons to fill their outermost level.

Chemical Bonding, continued

SECTION 2

Vocabulary

Read the following section highlights. Then in your own words, write the highlights in your ScienceLog.

1. ionic bond _____

_____

_____

_____

_____

_____

2. ions _____

_____

_____

_____

_____

_____

3. crystal lattice _____

_____

_____

_____

_____

_____

4. covalent bond _____

_____

_____

_____

_____

_____

CHAPTER 13

5. molecule _____

_____

_____

_____

_____

_____

6. metallic bond _____

_____

_____

_____

_____

_____

Notes

Read the following section highlights. Then in your own words, write the highlights in your ScienceLog.

• In ionic bonding, electrons are transferred between two atoms. The atom that loses electrons becomes a positive ion. The atom that gains electrons becomes a negative ion. The force of attraction between these oppositely charged ions is an ionic bond.

• Ionic bonding usually occurs between atoms of metals and atoms of nonmetals.

• Energy is needed to remove electrons from metal atoms to form positive ions. Energy is released when most nonmetal atoms gain electrons to form negative ions.

• In covalent bonding, electrons are shared by two atoms. The force of attraction between the nuclei of the atoms and the shared electrons is a covalent bond.

• Covalent bonding usually occurs between atoms of nonmetals.

• Electron-dot diagrams are a simple way to represent the valence electrons in an atom.

• Covalently bonded atoms form a particle called a molecule. A molecule is the smallest particle of a compound with the chemical properties of the compound.

• Diatomic elements are the only elements found in nature as diatomic molecules consisting of two atoms of the same element covalently bonded together.

• In metallic bonding, the outermost energy levels of metal atoms overlap, allowing the valence electrons to move throughout the metal. The force of attraction between a positive metal ion and the electrons in the metal is a metallic bond.

• Many properties of metals, such as conductivity, ductility, and malleability, result from the freely moving electrons in the metal.

CHAPTER

13 CHAPTER REVIEW WORKSHEET

Chemical Bonding

USING VOCABULARY

To complete the following sentences, choose the correct term from each pair of terms listed below, and write the term in the space provided.

1. The force of attraction that holds two atoms together is a

_____ . (crystal lattice or chemical bond)

2. Charged particles that form when atoms transfer electrons are

_____ . (molecules or ions)

3. The force of attraction between the nuclei of atoms and shared electrons is

a(n) _____ . (ionic bond or covalent bond)

4. Electrons free to move throughout a material are associated with a(n)

_____ . (ionic bond or metallic bond)

5. Shared electrons are associated with a _____ . (covalent bond or metallic bond)

UNDERSTANDING CONCEPTS

Multiple Choice

6. Which element has a full outermost energy level containing only two electrons?

   a. oxygen (O)   **c.** fluorine (F)
   b. hydrogen (H)   **d.** helium (He)

7. Which of the following describes what happens when an atom becomes an ion with a 2– charge?

   a. The atom gains 2 protons.
   b. The atom loses 2 protons.
   c. The atom gains 2 electrons.
   d. The atom loses 2 electrons.

8. The properties of ductility and malleability are associated with which type of bonds?

   a. ionic
   b. covalent
   c. metallic
   d. None of the above

9. In which area of the periodic table do you find elements whose atoms easily gain electrons?

   a. across the top two rows
   b. across the bottom row
   c. on the right side
   d. on the left side

▲ CHAPTER 13

Chemical Bonding, continued

10. What type of element tends to lose electrons when it forms bonds?

    a. metal

    b. metalloid

    c. nonmetal

    d. noble gas

11. Which pair of atoms can form an ionic bond?

    a. sodium (Na) and potassium (K)

    b. potassium (K) and fluorine (F)

    c. fluorine (F) and chlorine (Cl)

    d. sodium (Na) and neon (Ne)

Short Answer

12. List two properties of covalent compounds.

_____

_____

_____

13. Explain why an iron ion is attracted to a sulfide ion but not to a zinc ion.

_____

_____

_____

_____

_____

_____

_____

14. Using your knowledge of valence electrons, explain the main reason that so many different molecules are made from carbon atoms.

_____

_____

_____

_____

_____

Chemical Bonding, continued

---

15. Compare the three types of bonds based on what happens to the valence electrons of the atoms.

_____

_____

_____

_____

_____

CONCEPT MAPPING

16. Use the following terms to create a concept map: *chemical bonds, ionic bonds, covalent bonds, metallic bonds, molecule, ions*.

▲ **CHAPTER 13**

Chemical Bonding, continued

CRITICAL THINKING AND PROBLEM SOLVING

17. Predict the type of bond each of the following pairs of atoms would form:

   a. zinc (Zn) and zinc (Zn)

   _____

   b. oxygen (O) and nitrogen (N)

   _____

   c. phosphorus (P) and oxygen (O)

   _____

   d. magnesium (Mg) and chlorine (Cl)

   _____

18. Draw electron-dot diagrams for each of the following atoms, and state how many bonds it will have to make to fill its outer energy level.

   a. sulfur (S)

   _____

   b. nitrogen (N)

   _____

   c. neon (Ne)

   _____

d. iodine (I)

_____

e. silicon (Si)

_____

19. Does the substance being hit in the picture below contain ionic or metallic bonds? Explain.

_____

_____

_____

MATH IN SCIENCE

20. For each atom below, write the number of electrons it must gain or lose to have 8 valence electrons. Then calculate the charge of the ion that would form.

a. calcium (Ca)

_____

Chemical Bonding, continued

---

    b. phosphorus (P)

    _____

    c. bromine (Br)

    _____

    d. sulfur (S)

    _____

INTERPRETING GRAPHICS

Look at the picture of the wooden pencil below, and answer the following questions.

21. In which part of the pencil are metallic bonds found?

_____

_____

22. List three materials composed of molecules with covalent bonds.

_____

_____

23. Identify two differences between the properties of the metallically bonded material and one of the covalently bonded materials.

_____

_____

_____

_____

_____

NOW WHAT DO YOU THINK?

Take a minute to review your answers to the ScienceLog questions at the beginning of this chapter. Have your answers changed? If necessary, revise your answers based on what you have learned since you began this chapter. Record your revisions in your ScienceLog.

CHAPTER
14 **VOCABULARY & NOTES WORKSHEET**

Chemical Reactions

By studying the Vocabulary and Notes listed for each section below, you can gain a better understanding of this chapter.

SECTION 1

Vocabulary

In your own words, write a definition for each of the following terms in the space provided.

1. chemical reaction _____

_____

_____

2. chemical formula _____

_____

_____

3. chemical equation _____

_____

_____

4. reactants _____

_____

_____

5. products _____

_____

_____

CHAPTER 14

Chemical Reactions, continued

6. law of conservation of mass _____

_____

_____

_____

Notes

Read the following section highlights. Then, in your own words, write the highlights in your ScienceLog.

• Chemical reactions form new substances with different properties than the starting substances.

• Clues that a chemical reaction is taking place include formation of a gas or solid, a color change, and an energy change.

• A chemical formula tells the composition of a compound using chemical symbols and subscripts. Subscripts are small numbers written below and to the right of a symbol in a formula.

• Chemical formulas can sometimes be written from the names of covalent compounds and ionic compounds.

• A chemical equation describes a reaction using formulas, symbols, and coefficients.

• A balanced equation uses coefficients to illustrate the law of conservation of mass, which states that mass is neither created nor destroyed during a chemical reaction.

SECTION 2

Vocabulary

In your own words, write a definition for each of the following terms in the space provided.

1. synthesis reaction _____

_____

_____

_____

2. decomposition reaction _____

_____

_____

_____

3. single-replacement reaction _____

_____

_____

_____

4. double-replacement reaction _____

_____

_____

_____

Notes

- Many chemical reactions can be classified as one of four types by comparing reactants with products.
- In synthesis reactions, the reactants form a single product.
- In decomposition reactions, a single reactant breaks apart into two or more simpler products.
- In single-replacement reactions, a more-reactive element takes the place of a less-reactive element in a compound. No reaction will occur if a less-reactive element is placed with a compound containing a more-reactive element.
- In double-replacement reactions, ions in two compounds switch places. A gas or precipitate is often formed.

SECTION 3

Vocabulary

In your own words, write a definition for each of the following terms in the space provided.

1. exothermic _____

_____

_____

_____

2. endothermic _____

_____

_____

_____

CHAPTER 14

3. law of conservation of energy _____

_____

_____

_____

4. activation energy _____

_____

_____

_____

5. catalyst _____

_____

_____

_____

6. inhibitor _____

_____

_____

_____

Notes

Read the following section highlights. Then, in your own words, write the highlights in your ScienceLog.

- Energy is released in exothermic reactions. The energy released can be written as a product in a chemical equation.
- Energy is absorbed in endothermic reactions. The energy absorbed can be written as a reactant in a chemical equation.
- The law of conservation of energy states that energy is neither created nor destroyed.
- Activation energy is the energy needed to start a chemical reaction.
- Energy diagrams indicate whether a reaction is exothermic or endothermic by showing whether energy is given off or absorbed during the reaction.
- The rate of a chemical reaction is affected by temperature, concentration, surface area, and the presence of a catalyst or inhibitor.
- Raising the temperature, increasing the concentration, increasing the surface area, and adding a catalyst can increase the rate of a reaction.

CHAPTER

14   CHAPTER REVIEW WORKSHEET

Chemical Reactions

USING VOCABULARY

To complete the following sentences, choose the correct term from each pair of terms listed below, and write the term in the space provided.

1. Adding a(n) _____ will slow down a chemical reaction. (catalyst or inhibitor)

2. A chemical reaction that gives off light is called _____. (exothermic or endothermic)

3. A chemical reaction that forms one compound from two or more substances is called

a _____. (synthesis reaction or decomposition reaction)

4. The 2 in the formula Ag_2S is a _____. (subscript or coefficient)

5. The starting materials in a chemical reaction are _____. (reactants or products)

UNDERSTANDING CONCEPTS

Multiple Choice

6. Balancing a chemical equation so that the same number of atoms of each element is found in both the reactants and the products is an illustration of

  a. activation energy.

  b. the law of conservation of energy.

  c. the law of conservation of mass.

  d. a double-replacement reaction.

7. What is the correct chemical formula for calcium chloride?

  a. $CaCl$ **c.** Ca_2Cl

  b. $CaCl_2$ **d.** Ca_2Cl_2

8. In which type of reaction do ions in two compounds switch places?

  a. synthesis

  b. decomposition

  c. single-replacement

  d. double-replacement

9. Which is an example of the use of activation energy?

  a. plugging in an iron

  b. playing basketball

  c. holding a lit match to paper

  d. eating

Chemical Reactions, continued

10. Enzymes in your body act as catalysts. Thus, the role of enzymes is to
    a. increase the rate of chemical reactions.
    b. decrease the rate of chemical reactions.
    c. help you breathe.
    d. inhibit chemical reactions.

Short Answer

11. Classify each of the following reactions:
    a. $Fe + O_2 \rightarrow Fe_2O_3$

    _____

    _____

    b. $Al + CuSO_4 \rightarrow Al_2(SO_4)_3 + Cu$

    _____

    _____

    c. $Ba(CN)_2 + H_2SO_4 \rightarrow BaSO_4 + HCN$

    _____

    _____

12. Name two ways that you could increase the rate of a chemical reaction.

    _____

    _____

    _____

    _____

13. Acetic acid, a compound found in vinegar, reacts with baking soda to produce carbon dioxide, water, and sodium acetate. Without writing an equation, identify the reactants and the products of this reaction.

    _____

    _____

    _____

    _____

Chemical Reactions, continued

CONCEPT MAPPING

14. Use the following terms to create a concept map: *chemical reaction, chemical equation, chemical formulas, reactants, products, coefficients, subscripts.*

_____

Chemical Reactions, continued

CRITICAL THINKING AND PROBLEM SOLVING

15. Your friend is very worried by rumors he has heard about a substance called dihydrogen monoxide. What could you say to your friend to calm his fears? (Be sure to write the formula of the substance.)

_____

_____

16. As long as proper safety precautions have been taken, how can explosives be transported long distances without exploding?

_____

_____

_____

_____

MATH IN SCIENCE

17. Calculate the number of atoms of each element shown in each of the following:

  a. $CaSO_4$

_____

_____

  b. $4NaOCl$

_____

_____

  c. $Fe(NO_3)_2$

_____

_____

  d. $2Al_2(CO_3)_3$

_____

_____

Chemical Reactions, continued

18. Write balanced equations for the following:

a. $Fe + O_2 \rightarrow Fe_2O_3$

_____

_____

_____

b. $Al + CuSO_4 \rightarrow Al_2(SO_4)_3 + Cu$

_____

_____

_____

c. $Ba(CN)_2 + H_2SO_4 \rightarrow BaSO_4 + HCN$

_____

_____

_____

19. Write and balance chemical equations from each of the following descriptions:

a. Bromine reacts with sodium iodide to form iodine and sodium bromide.

_____

_____

b. Phosphorus reacts with oxygen gas to form diphosphorus pentoxide.

_____

_____

c. Lithium oxide decomposes to form lithium and oxygen.

_____

_____

CHAPTER 14

INTERPRETING GRAPHICS

20. Look at the picture on page 369 in the top right side of the page. What evidence in the picture supports the claim that a chemical reaction is taking place?

_____

_____

21. Use the energy diagram below to answer the questions that follow.

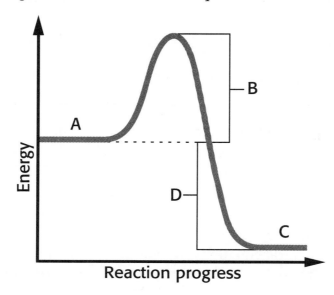

   a. Which letter represents the energy of the products?

   _____

   b. Which letter represents the activation energy of the reaction?

   _____

   c. Is energy given off or absorbed by this reaction?

   _____

NOW WHAT DO YOU THINK?

Take a minute to review your answers to the ScienceLog questions at the beginning of the chapter. Have your answers changed? If necessary, revise your answers based on what you have learned since you began this chapter. Record your revisions in your ScienceLog.

CHAPTER

15 **VOCABULARY & NOTES WORKSHEET**

Chemical Compounds

By studying the Vocabulary and Notes listed for each section below, you can gain a better understanding of this chapter.

SECTION 1

Vocabulary

In your own words, write a definition for each of the following terms in the space provided.

1. ionic compounds _____

_____

_____

2. covalent compounds _____

_____

_____

Notes

Read the following section highlights. Then, in your own words, write the highlights in your ScienceLog.

• Ionic compounds contain ionic bonds and are composed of oppositely charged ions arranged in a repeating pattern called a crystal lattice.

• Ionic compounds tend to be brittle, have high melting points, and dissolve in water to form solutions that conduct an electric current.

• Covalent compounds are composed of elements that are covalently bonded and consist of independent particles called molecules.

• Covalent compounds tend to have low melting points. Most do not dissolve well in water and do not form solutions that conduct an electric current.

SECTION 2

Vocabulary

In your own words, write a definition for each of the following terms in the space provided.

1. acid _____

_____

_____

_____

CHAPTER 15

Chemical Compounds, continued

2. base _____

_____

_____

_____

_____

3. pH _____

_____

_____

_____

_____

_____

_____

4. salt _____

_____

_____

_____

_____

_____

Notes

Read the following section highlights. Then, in your own words, write the highlights in your ScienceLog.

- An acid is a compound that increases the number of hydrogen ions in solution. Acids taste sour, turn blue litmus paper red, react with metals to produce hydrogen gas, and react with limestone or baking soda to produce carbon dioxide gas.

- A base is a compound that increases the number of hydroxide ions in solution. Bases taste bitter, feel slippery, and turn red litmus paper blue.

- When dissolved in water, every molecule of a strong acid or base breaks apart to form ions. Few molecules of weak acids and bases break apart to form ions.

- When combined, an acid and a base neutralize one another to produce water and a salt.

- pH is a measure of hydronium ion concentration in a solution. A pH of 7 indicates a neutral substance. A pH of less than 7 indicates an acidic substance. A pH of greater than 7 indicates a basic substance.

- A salt is an ionic compound formed from the positive ion of a base and the negative ion of an acid.

SECTION 3

Vocabulary

In your own words, write a definition for each of the following terms in the space provided.

1. organic compounds _____

_____

_____

_____

_____

_____

2. biochemicals _____

_____

_____

_____

_____

_____

3. carbohydrates _____

_____

_____

_____

_____

_____

4. lipids _____

_____

_____

_____

CHAPTER 15

---

5. proteins _____

_____

_____

_____

6. nucleic acids _____

_____

_____

7. hydrocarbons _____

_____

_____

_____

Notes

Read the following section highlights. Then, in your own words, write the highlights in your ScienceLog.

- Organic compounds are covalent compounds composed of carbon-based molecules.
- Each carbon atom forms four bonds with other carbon atoms or with atoms of other elements to form straight chains, branched chains, or rings.
- Biochemicals are organic compounds made by living things.
- Carbohydrates are biochemicals that are composed of one or more simple sugars bonded together; they are used as a source of energy and for energy storage.
- Lipids are biochemicals that do not dissolve in water and have many functions, including storing energy and making up cell membranes.
- Proteins are biochemicals that are composed of amino acids and have many functions, including regulating chemical activities, transporting and storing materials, and providing structural support.
- Nucleic acids are biochemicals that store information and help to build proteins and other nucleic acids.
- Hydrocarbons are organic compounds composed of only carbon and hydrogen.
- In a saturated hydrocarbon, each carbon atom in the molecule shares a single bond with each of four other atoms.
- In an unsaturated hydrocarbon, at least two carbon atoms share a double bond or a triple bond.
- Many aromatic hydrocarbons are based on the six-carbon ring of benzene.
- Other organic compounds, including alkyl halides, alcohols, organic acids, and esters, are formed by adding atoms of other elements.

CHAPTER

15 CHAPTER REVIEW WORKSHEET

Chemical Compounds

USING VOCABULARY

To complete the following sentences, choose the correct term from each pair of terms listed below, and write the term in the space provided.

1. Compounds that have low melting points and do not usually dissolve well in water are _____ . (ionic compounds or covalent compounds)

2. A(n) _____ turns red litmus paper blue. (acid or base)

3. _____ are composed of only carbon and hydrogen. (Ionic compounds or Hydrocarbons)

4. A biochemical composed of amino acids is a _____ . (lipid or protein)

5. A source of energy for living things can be found in _____ . (nucleic acids or carbohydrates)

UNDERSTANDING CONCEPTS

Multiple Choice

6. Which of the following describes lipids?
   a. used to store energy
   b. do not dissolve in water
   c. make up most of the cell membrane
   d. All of the above

7. An acid reacts to produce carbon dioxide when the acid is added to
   a. water.
   b. limestone.
   c. salt.
   d. sodium hydroxide.

8. Which of the following does NOT describe ionic compounds?
   a. high melting point
   b. brittle
   c. do not conduct electric currents in water
   d. dissolve easily in water

9. An increase in the amount of hydrogen ions in solution _____ the pH.
   a. raises
   b. lowers
   c. does not affect
   d. doubles

10. Which of the following compounds makes up the majority of cell membranes?
   a. lipids
   b. ionic compounds
   c. acids
   d. nucleic acids

CHAPTER 15

11. The compounds that store information for building proteins are

   a. lipids.

   b. hydrocarbons.

   c. nucleic acids.

   d. carbohydrates.

Short Answer

12. What type of compound would you use to neutralize a solution of potassium hydroxide?

_____

_____

_____

_____

13. Explain why the reaction of an acid with a base is called *neutralization*.

_____

_____

_____

_____

_____

14. What characteristic of carbon atoms helps to explain the wide variety of organic compounds?

_____

_____

_____

_____

_____

15. Compare acids and bases based on the ion produced when each compound is dissolved in water.

_____

_____

_____

Chemical Compounds, continued

CONCEPT MAPPING

16. Use the following terms to create a concept map: _acid, base, salt, neutral, pH._

CRITICAL THINKING AND PROBLEM SOLVING

17. Fish give off the base ammonia, NH_3, as waste. How does the release of ammonia affect the pH of the water in the aquarium? What can be done to correct the problem?

_____

_____

_____

_____

Chemical Compounds, continued

_____

18. Many insects, such as fire ants, inject formic acid, a weak acid, when they bite or sting. Describe the type of compound that should be used to treat the bite.

_____

_____

_____

_____

_____

19. Organic compounds are also covalent compounds. What properties would you expect organic compounds to have as a result?

_____

_____

_____

_____

_____

20. Farmers often can taste their soil to determine whether the soil has the correct acidity for their plants. How would taste help the farmer determine the acidity of the soil?

_____

_____

_____

_____

21. A diet that includes a high level of lipids is unhealthy. Why is a diet containing no lipids also unhealthy?

_____

_____

_____

_____

INTERPRETING GRAPHICS

Study the structural formulas below, and then answer the questions that follow.

22. A saturated hydrocarbon is represented by which structural formula(s)?

_____

23. An unsaturated hydrocarbon is represented by which structural formula(s)?

_____

24. An aromatic hydrocarbon is represented by which structural formula(s)?

_____

NOW WHAT DO YOU THINK?

Take a minute to review your answers to the ScienceLog questions at the beginning of the chapter. Have your answers changed? If necessary, revise your answers based on what you have learned since you began this chapter. Record your revisions in your ScienceLog.

CHAPTER

16 VOCABULARY & NOTES WORKSHEET

Atomic Energy

By studying the Vocabulary and Notes listed for each section below, you can gain a better understanding of this chapter.

SECTION 1

Vocabulary

In your own words, write a definition for the following terms in the space provided.

1. nuclear radiation _____

_____

_____

_____

2. radioactivity _____

_____

_____

_____

3. radioactive decay _____

_____

_____

_____

4. alpha decay _____

5. mass number _____

_____

_____

_____

6. beta decay _____

_____

_____

_____

Atomic Energy, continued

7. isotopes _____

_____

8. gamma decay _____

_____

_____

_____

9. half-life _____

_____

Notes

Read the following section highlights. Then, in your own words, write the highlights in your ScienceLog.

- Radioactive nuclei give off nuclear radiation in the form of alpha particles, beta particles, and gamma rays through a process called radioactive decay.
- During alpha decay, an alpha particle is released from the nucleus. An alpha particle is composed of two protons and two neutrons.
- During beta decay, a beta particle is released from the nucleus. A beta particle can be an electron or a positron.
- Gamma decay occurs with alpha decay and beta decay when particles in the nucleus rearrange and emit energy in the form of gamma rays.
- Gamma rays penetrate matter better than alpha or beta particles. Beta particles penetrate matter better than alpha particles.
- Nuclear radiation can damage living and nonliving matter.
- Half-life is the amount of time it takes for one-half of the nuclei of a radioactive isotope to decay. The age of some objects can be determined using half-lives.
- Uses of radioactive materials include detecting defects in materials, sterilizing products, tracing a plant's or animal's use of an element, diagnosing illness, and producing electrical energy.

SECTION 2

Vocabulary

In your own words, write a definition for the following terms in the space provided.

1. nuclear fission _____

_____

_____

2. nuclear chain reaction _____

_____

_____

_____

3. nuclear fusion _____

_____

_____

_____

Notes

Read the following section highlights. Then, in your own words, write the highlights in your ScienceLog.

- Nuclear fission occurs when a massive, unstable nucleus breaks into two less massive nuclei. Nuclear fission is used in power plants to generate electrical energy.
- Nuclear fusion occurs when two or more nuclei combine to form a larger nucleus. The sun's energy comes from the fusion of hydrogen to form helium.
- The energy released by nuclear fission and nuclear fusion is produced when matter is converted into energy.
- Nuclear power plants use nuclear fission to supply many homes with electrical energy without releasing carbon dioxide or other gases into the atmosphere. A limited fuel supply, radioactive waste products, and the possible release of radioactive material are disadvantages of fission.
- Fuel for nuclear fusion is plentiful, and only small amounts of radioactive waste products are produced. Fusion is not currently a practical energy source because of the large amount of energy needed to heat and contain the hydrogen plasma.

CHAPTER

16 **CHAPTER REVIEW WORKSHEET**

Atomic Energy

USING VOCABULARY

The statements below are false. For each statement, replace the underlined term to make a true statement and write the correct term in the space provided.

1. <u>Nuclear fusion</u> involves splitting a nucleus.

_____

2. During one <u>beta decay,</u> half of a radioactive sample will decay.

_____

3. <u>Nuclear fission</u> includes the particles and rays released by radioactive nuclei.

_____

4. <u>Alpha decay</u> occurs during the rearrangement of protons and neutrons in the nucleus.

_____

UNDERSTANDING CONCEPTS

Multiple Choice

5. Which of the following is a use of radioactive material?

    a. detecting smoke
    b. locating defects in materials
    c. generating electrical energy
    d. All of the above

6. Which particle both begins and is produced by a nuclear chain reaction?

    a. positron   **c.** alpha particle
    b. neutron   **d.** beta particle

7. Nuclear radiation that can be stopped by paper is called

    a. alpha particles.   **c.** gamma rays.
    b. beta particles.   **d.** None of the above

8. The half-life of a radioactive atom is 2 months. If you start with 1g of the element, how much will remain after 6 months?

    a. One-half of a gram will remain.
    b. One-fourth of a gram will remain.
    c. One-eighth of a gram will remain.
    d. None will remain.

9. The waste products of nuclear fission

    a. are harmless.
    b. are safe after 20 years.
    c. can be destroyed by burning them.
    d. remain radioactive for thousands of years.

Atomic Energy, continued

10. Which statement about nuclear fusion is false?
   a. Nuclear fusion occurs in the sun.
   b. Nuclear fusion is the joining of the nuclei of atoms.
   c. Nuclear fusion is currently used to generate electrical energy.
   d. Nuclear fusion uses hydrogen as fuel.

Short Answer

11. What conditions could cause a nucleus to be unstable?

_____

_____

_____

_____

12. What are two dangers associated with nuclear fission?

_____

_____

_____

_____

13. What are two of the problems that need to be solved in order to make nuclear fusion a practical energy source?

_____

_____

_____

_____

_____

14. In fission, the products have less mass than the starting materials. Explain what happened.

_____

_____

_____

Atomic Energy, continued

CONCEPT MAPPING

15. Use the following terms to create a concept map: *radioactive decay, alpha particle, beta particle, gamma ray, nuclear radiation.*

_____

Atomic Energy, continued

_____

CRITICAL THINKING AND PROBLEM SOLVING

Write one or two sentences to answer the following questions:

16. Smoke detectors often use americium-243 to detect smoke particles in the air. Americium-243 undergoes alpha decay. Do you think that these smoke detectors are safe to have in your home if used properly? Explain. (Hint: How penetrating are alpha particles?)

_____

_____

_____

_____

_____

17. Explain how radiation can cause cancer.

_____

_____

_____

_____

18. Explain why nuclei of carbon, oxygen, and even iron can be found in stars.

_____

_____

_____

_____

_____

_____

_____

19. If you could block all radiation from sources outside your body, explain why you would still be exposed to some radiation.

_____

_____

_____

_____

_____

_____

20. A scientist used 10 g of phosphorus-32 in a test on plant growth but forgot to record the date. When he measured the phosphorus-32 some time later, he found only 2.5 g remaining. If the half-life is 14 days, how many days ago did he start the experiment?

_____

_____

_____

_____

INTERPRETING GRAPHICS

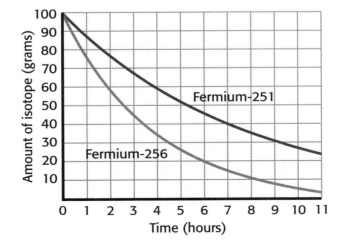

Atomic Energy, continued

21. Use the graph on the previous page to answer the questions below:
    a. What is the half-life of fermium-256? of fermium-251?

_____

_____

    b. Which of these isotopes is more stable? Explain.

_____

_____

22. The image of a small purse, shown below, was made in a similar manner as Becquerel's original experiment. What conclusions can be drawn about the penetrating power of radiation from this image?

_____

_____

_____

_____

_____

NOW WHAT DO YOU THINK?

Take a minute to review your answers to the ScienceLog questions at the beginning of the chapter. Have your answers changed? If necessary, revise your answers based on what you have learned since you began this chapter. Record your revisions in your ScienceLog.

Name _____ Date _____ Class_____

Introduction to Electricity

By studying the Vocabulary and Notes listed for each section below, you can gain a better understanding of this chapter.

SECTION 1

Vocabulary

In your own words, write a definition for each of the following terms in the space provided.

1. law of electric charges _____

_____

2. electric force _____

_____

3. conduction _____

_____

_____

_____

4. induction _____

_____

_____

_____

5. conductor _____

_____

6. insulator _____

_____

7. static electricity _____

_____

8. electric discharge _____

_____

Notes

Read the following section highlights. Then, in your own words, write the highlights in your ScienceLog.

- The law of electric charges states that like charges repel and opposite charges attract.
- The electric force varies depending on the size of the charges exerting the force and the distance between them.
- Objects become charged when they gain or lose electrons. Objects may become charged by friction, conduction, or induction.
- Charges are not created or destroyed and are said to be conserved.
- An electroscope can be used to detect charges.
- Charges move easily in conductors but do not move easily in insulators.
- Static electricity is the buildup of electric charges on an object. Static electricity is lost through electric discharge. Lightning is a form of electric discharge.
- Lightning rods work by grounding the electric charge carried by lightning.

SECTION 2

Vocabulary

In your own words, write a definition for each of the following terms in the space provided.

1. cell _____

_____

2. battery _____

_____

_____

3. potential difference _____

_____

_____

4. photocell _____

_____

_____

5. thermocouple _____

_____

_____

Introduction to Electricity, continued

Notes

- Batteries are made of cells that convert chemical energy to electrical energy.
- Electric currents can be produced when there is a potential difference.
- Photocells and thermocouples are devices used to produce electrical energy.

SECTION 3

Vocabulary

In your own words, write a definition for each of the following terms in the space provided.

1. current _____

_____

_____

2. voltage _____

_____

_____

3. resistance _____

_____

_____

4. electric power _____

_____

Notes

Read the following section highlights. Then, in your own words, write the highlights in your ScienceLog.

- Electrical current is a continuous flow of charge caused by the motion of electrons.
- Voltage is the same as potential difference. As voltage increases, current increases.
- An object's resistance varies depending on the object's material, thickness, length, and temperature. As resistance increases, current decreases.
- Ohm's law describes the relationship between current, resistance, and voltage.
- Electric power is the rate at which electrical energy does work. It is expressed in watts or kilowatts.
- Electrical energy is electric power multiplied by time. It is usually expressed in kilowatt-hours.

SECTION 4

Vocabulary

In your own words, write a definition for each of the following terms in the space provided.

1. circuit _____

_____

2. load _____

_____

3. series circuit _____

_____

4. parallel circuit _____

_____

Notes

Read the following section highlights. Then, in your own words, write the highlights in your ScienceLog.

• Circuits consist of an energy source, a load, wires, and sometimes a switch.

• All parts of a series circuit are connected in a single loop.

• The loads in a parallel circuit are on separate branches.

• Circuits can fail because of a short circuit or circuit overload.

• Fuses or circuit breakers protect your home against circuit failure.

CHAPTER

17 CHAPTER REVIEW WORKSHEET

Introduction to Electricity

USING VOCABULARY

To complete the following sentences, choose the correct term from each pair of terms listed below, and write the term in the space provided.

1. A _____ converts chemical energy into electrical energy. (battery or photocell)

2. Charges flow easily in a(n) _____ . (insulator or conductor)

3. _____ is the opposition to the flow of electric charge. (Resistance or Electric power)

4. A _____ is a complete, closed path through which charges flow. (load or circuit)

5. Lightning is a form of _____ . (static electricity or electric discharge)

UNDERSTANDING CONCEPTS

Multiple Choice

6. If two charges repel each other, the two charges must be
   a. positive and positive.
   b. positive and negative.
   c. negative and negative.
   d. Either (a) or (c)

7. A device that can convert chemical energy to electrical energy is a
   a. lightning rod.
   b. cell.
   c. light bulb.
   d. All of the above

8. Which of the following wires has the lowest resistance?
   a. a short, thick copper wire at 25°C
   b. a long, thick copper wire at 35°C
   c. a long, thin copper wire at 35°C
   d. a short, thick iron wire at 25°C

9. An object becomes charged when the atoms in the object gain or lose
   a. protons.   **c.** electrons.
   b. neutrons.   **d.** All of the above

10. A device used to protect buildings from electrical fires is a(n)
   a. electric meter.   **c.** fuse.
   b. circuit breaker.   **d.** Both (b) and (c)

11. In order to produce a current from a cell, the electrodes of the cell must
   a. have a potential difference.
   b. be in a liquid.
   c. be exposed to light.
   d. be at two different temperatures.

12. What type of current comes from the outlets in your home?
   a. direct current
   b. alternating current
   c. electric discharge
   d. static electricity

Short Answer

13. List and describe the three essential parts of a circuit.

_____

_____

_____

_____

_____

_____

14. Name the two factors that affect the strength of electric force, and explain how they affect electric force.

_____

_____

_____

_____

_____

15. Describe how direct current differs from alternating current.

_____

_____

_____

_____

Introduction to Electricity, continued

CONCEPT MAPPING

16. Use the following terms to create a concept map: *electric current, battery, charges, photocell, thermocouple, circuit, parallel circuit, series circuit.*

CHAPTER 17

CRITICAL THINKING AND PROBLEM SOLVING

17. Your science classroom was rewired over the weekend. On Monday, you notice that the electrician may have made a mistake. In order for the fish-tank bubbler to work, the lights in the room must be on. And if you want to use the computer, you must turn on the overhead projector. Describe what mistake the electrician made with the circuits in your classroom.

_____

_____

_____

_____

18. You can make a cell using an apple, a strip of copper, and a strip of silver. Explain how you would construct the cell, and identify the parts of the cell. What type of cell is formed? Explain your answer.

_____

_____

_____

_____

_____

_____

19. Your friend shows you a magic trick. She rubs a plastic comb with a piece of silk and holds the comb close to a stream of water. When the comb is close to the water, the water bends toward the comb. Explain how this trick works. (Hint: Think about how objects become charged.)

_____

_____

_____

_____

_____

_____

_____

_____

MATH IN SCIENCE

Use Ohm's law to solve the following problems:

20. What voltage is needed to produce a 6 A current through a resistance of 3 Ω?

_____

21. Find the current produced when a voltage of 60 V is applied to a resistance of 15 Ω.

_____

22. What is the resistance of an object if a voltage of 40 V produces a current of 5 A?

_____

INTERPRETING GRAPHICS

23. Classify the objects in the illustration below as conductors or insulators.

_____

_____

NOW WHAT DO YOU THINK?

Take a minute to review your answers to the ScienceLog questions at the beginning of this chapter. Have your answers changed? If necessary, revise your answers based on what you have learned since you began this chapter. Record your revisions in your ScienceLog.

CHAPTER

18 VOCABULARY & NOTES WORKSHEET

Electromagnetism

SECTION 1

Vocabulary

In your own words, write a definition for the following terms in the space provided.

1. magnet _____

_____

2. poles _____

_____

3. magnetic force _____

_____

Notes

Read the following section highlights. Then, in your own words, write the highlights in your ScienceLog.

- All magnets have two poles. One pole will always point to the north if allowed to rotate freely, and it is called the north pole. The other pole is called the south pole.

- Like magnetic poles repel each other; opposite magnetic poles attract.

- All magnets are surrounded by a magnetic field. The shape of that magnetic field can be shown with magnetic field lines.

- A material is magnetic if its domains are aligned. Iron, nickel, and cobalt atoms group together in domains.

- Magnets can be classified as ferromagnets, electromagnets, temporary magnets, and permanent magnets. A magnet can belong to more than one group.

- Earth acts as if it has a big bar magnet in its core.

- Compass needles and the north pole of magnets point to Earth's magnetic south pole— which is close to Earth's geographic North Pole.

- Auroras are most commonly seen near Earth's magnetic poles because Earth's magnetic fields bend inward at the poles.

SECTION 2

Vocabulary

In your own words, write a definition for the following terms in the space provided.

1. electromagnetism _____

_____

2. solenoid _____

_____

Electromagnetism, continued

3. electromagnet _____

_____

4. electric motor _____

_____

Notes

Read the following section highlights. Then, in your own words, write the highlights in your ScienceLog.

- Oersted discovered that a wire carrying an electric current produces a magnetic field.
- Electromagnetism is the interaction between electricity and magnetism.
- A solenoid is a coil of current-carrying wire that produces a magnetic field.
- An electromagnet is a solenoid with an iron core. The electromagnet has a stronger magnetic field than a solenoid of the same size does.
- Increasing the current in a solenoid or an electromagnet increases the magnetic field.
- Increasing the number of loops on a solenoid or an electromagnet increases the magnetic field.
- A magnet can exert a force on a wire carrying a current.
- In a doorbell, the magnetic field of a solenoid pulls an iron rod, and the iron rod strikes the bell.
- The magnetic force between a magnet and wires carrying an electric current makes an electric motor turn.
- An electric motor converts electrical energy into kinetic energy.
- A galvanometer measures current by using the magnetic force between an electromagnet and a permanent magnet.

SECTION 3

Vocabulary

In your own words, write a definition for the following terms in the space provided.

1. electromagnetic induction _____

_____

2. generator _____

_____

3. transformer _____

_____

Notes

Read the following section highlights. Then, in your own words, write the highlights in your ScienceLog.

• Faraday discovered that a changing magnetic field can create an electric current in a wire. This is called electromagnetic induction.

• Generators use electromagnetic induction to convert kinetic energy into electrical energy.

• Kinetic energy can be supplied to a generator in different ways.

• Transformers increase or decrease the voltage of an alternating current using electromagnetic induction.

• A step-up transformer increases the voltage of an alternating current. Its primary coil has fewer loops than its secondary coil.

• A step-down transformer decreases the voltage of an alternating current. Its primary coil has more loops than its secondary coil.

CHAPTER

18 CHAPTER REVIEW WORKSHEET

Electromagnetism

To complete the following sentences, choose the correct term from each pair of terms listed below, and write the term in the space provided.

1. All magnets have two _____. (magnetic forces or poles)

2. A(n) _____ converts kinetic energy into electrical energy. (electric motor or generator)

3. _____ occurs when an electric current is produced by a changing magnetic field. (Electromagnetic induction or Magnetic force)

4. The interaction between electricity and magnetism is called

   _____. (electromagnetism or electromagnetic induction)

Multiple Choice

5. The region around a magnet in which magnetic forces can act is called the
   a. magnetic field.
   b. domain.
   c. pole.
   d. solenoid.

6. An electric fan has an electric motor inside to change
   a. kinetic energy into electrical energy.
   b. thermal energy into electrical energy.
   c. electrical energy into thermal energy.
   d. electrical energy into kinetic energy.

7. The marked end of a compass needle always points directly to
   a. Earth's geographic South Pole.
   b. Earth's geographic North Pole.
   c. a magnet's south pole.
   d. a magnet's north pole.

8. A device that increases the voltage of an alternating current is called a(n)
   a. electric motor.
   b. galvanometer.
   c. step-up transformer.
   d. step-down transformer.

9. The magnetic field of a solenoid can be increased by
   a. adding more loops.
   b. increasing the current.
   c. putting an iron core inside the coil to make an electromagnet.
   d. All of the above

10. What do you end up with if you cut a magnet in half?
    a. one north-pole piece and one south-pole piece
    b. two unmagnetized pieces
    c. two pieces, each with a north pole and a south pole
    d. two north-pole pieces

CHAPTER 18 ▲ ▲ ▲

Short Answer

11. Explain why auroras are seen mostly near the North and South Poles.

_____

_____

_____

_____

_____

_____

12. Compare the function of a generator with the function of an electric motor.

_____

_____

_____

_____

_____

_____

13. Explain why some pieces of iron are more magnetic than others.

_____

_____

_____

_____

_____

_____

Electromagnetism, continued

CONCEPT MAPPING

14. Use the following terms to create a concept map: *electromagnetism, electricity, magnetism, electromagnetic induction, generators, transformers.*

▲▲ CHAPTER 18

CRITICAL THINKING AND PROBLEM SOLVING

Write one or two sentences to answer the following questions:

15. You win a hand-powered flashlight as a prize in your school science fair. The flashlight has a clear plastic case so you can look inside to see how it works. When you press the handle, a gray ring spins between two coils of wire. The ends of the wire are connected to the light bulb. So when you press the handle, the light bulb glows. Explain how an electric current is produced to light the bulb. (Hint: Paper clips are attracted to the gray ring.)

_____

_____

_____

_____

_____

_____

_____

16. Fire doors are doors that can slow the spread of fire from room to room when they are closed. In some buildings, fire doors are held open by electromagnets. The electromagnets are controlled by the building's fire alarm system. If a fire is detected, the doors automatically shut. Explain why electromagnets are used instead of permanent magnets.

_____

_____

_____

_____

_____

_____

_____

Electromagnetism, continued

INTERPRETING GRAPHICS

17. Study the solenoids and electromagnets shown below. Rank them in order of strongest magnetic field to weakest magnetic field. Explain your ranking.

a

Current = 2 A

c

Current = 4 A

b

Current = 2 A

d

Current = 4 A

_____

_____

_____

_____

_____

_____

_____

_____

_____

NOW WHAT DO YOU THINK?

Take a minute to review your answers to the ScienceLog questions at the beginning of the chapter. Have your answers changed? If necessary, revise your answers based on what you have learned since you began this chapter. Record your revisions in your ScienceLog.

▲▶ **CHAPTER 18**

CHAPTER

19 VOCABULARY & NOTES WORKSHEET

Electronic Technology

By studying the Vocabulary and Notes listed for each section below, you can gain a better understanding of this chapter.

SECTION 1

Vocabulary

In your own words, write a definition for the following terms in the space provided.

1. circuit board _____

_____

_____

_____

2. semiconductor _____

_____

_____

3. doping _____

_____

_____

_____

_____

4. diode _____

_____

5. transistor _____

_____

6. integrated circuit _____

_____

_____

Electronic Technology, continued

Notes

Read the following section highlights. Then, in your own words, write the highlights in your ScienceLog.

- Electronic devices use electrical energy to transmit information.
- Many electronic components are made of semiconductors. Two types of semiconductors result from a process called doping. They are n-type and p-type semiconductors.
- The two types of semiconductors can be sandwiched together to produce diodes and transistors.
- Diodes allow electric current in only one direction.
- Transistors can be used as amplifiers or as switches.
- Integrated circuits can contain many electronic components. They allow electronic systems to be smaller.

SECTION 2

Vocabulary

In your own words, write a definition for the following terms in the space provided.

1. telecommunication _____

_____

2. signal _____

_____

3. analog signal _____

_____

_____

_____

4. digital signal _____

_____

Notes

Read the following section highlights. Then, in your own words, write the highlights in your ScienceLog.

- Electronic devices use signals to transmit information. The signals are usually contained in another form of energy, such as radio waves or electric current.
- The properties of analog signals change continuously according to changes in the original signal. Telephones use analog signals.

▲ **CHAPTER 19**
▲
▲

- A digital signal is a series of electrical pulses that represents the digits of binary numbers. CD players use digital signals.
- Sound can be recorded digitally or as an analog signal.
- Radio and television rely on electromagnetic waves.
- In radio, signals that represent sound are combined with radio waves and sent through the air. Radios can pick up the radio waves and convert them back to sound waves.
- A color television image is produced by three electron beams that scan the screen of a cathode-ray tube, or CRT. Fluorescent materials on the screen glow to create the picture.

SECTION 3

Vocabulary

In your own words, write a definition for the following terms in the space provided.

1. computer _____

_____

2. microprocessor _____

_____

3. hardware _____

4. software _____

_____

5. Internet _____

_____

_____

Notes

Read the following section highlights. Then, in your own words, write the highlights in your ScienceLog.

- The basic functions of a computer involve input, processing, memory, and output. A computer cannot perform a task without a set of commands.
- The first computers were very large and could not perform many tasks.
- Because microprocessors contain many computer capabilities on a single chip, computers have been reduced in size.
- Computer hardware refers to the parts or the equipment that make up a computer.
- Computer software is a set of instructions or commands that tells a computer what to do.
- Modems allow millions of computers to connect with one another and share information on the Internet.

CHAPTER

19 CHAPTER REVIEW WORKSHEET

Electronic Technology

USING VOCABULARY

For each pair of terms, explain the difference in their meanings.

1. semiconductor/doping _____

_____

_____

_____

2. transistor/diode _____

_____

_____

_____

_____

3. signal/telecommunication _____

_____

_____

4. analog signal/digital signal _____

_____

_____

_____

5. computer/microprocessor _____

_____

_____

_____

_____

6. hardware/software _____

_____

_____

Electronic Technology, continued

Multiple Choice

7. All electronic devices transmit information using

   a. signals.

   b. electromagnetic waves.

   c. radio waves.

   d. modems.

8. Semiconductors are used to make

   a. transistors.

   b. integrated circuits.

   c. diodes.

   d. All of the above

9. Which of the following is an example of a telecommunication device?

   a. vacuum tube

   b. telephone

   c. radio

   d. Both (b) and (c)

10. A monitor, printer, and speaker are examples of

   a. input devices.

   b. memory.

   c. computers.

   d. output devices.

11. Record players play sounds that were recorded in the form of

   a. digital signals.

   b. electric current.

   c. analog signals.

   d. radio waves.

12. Memory in a computer that is permanent and cannot be added to is called

   a. RAM.

   b. ROM.

   c. CPU.

   d. None of the above

13. Cathode-ray tubes are used in

   a. telephones.

   b. telegraphs.

   c. televisions.

   d. radios.

Short Answer

14. How is an electronic device different from a machine that uses electrical energy?

_____

_____

_____

_____

15. How does a diode allow current to flow in one direction?

_____

_____

_____

_____

Electronic Technology, continued

16. In one or two sentences, describe how a TV works.

_____

_____

_____

_____

17. Give three examples of how computers are used in your everyday life.

_____

_____

_____

_____

18. Explain the advantages that transistors have over vacuum tubes.

_____

_____

CONCEPT MAPPING

19. Use the following terms to create a concept map: *electronic devices, radio waves, electric current, signals, information.*

▲▲ CHAPTER 19
▲

20. Your friend is preparing an oral report on the history of radio and finds a photograph of a large radio from the 1940s. "Why is this radio so huge?" he asks you. Using what you know about electronic devices, how do you explain the size of this vintage radio?

_____

_____

_____

_____

21. Using what you know about the differences between analog and digital signals, explain how the sound from a record player is different from the sound from a CD player.

_____

_____

_____

_____

22. What do Morse code and digital signals have in common?

_____

_____

_____

23. Computers can process a lot of information, but they cannot think. Explain why this is true.

_____

_____

_____

24. Based on what you learned in the chapter, how do you think an automatic garage door opener works?

_____

_____

_____

_____

Look at the diagram below, and answer the questions that follow.

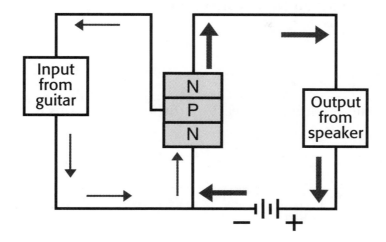

25. What purpose does the transistor serve in this situation?

_____

_____

26. How does the current in the left side of the circuit compare with the current in the right side of the circuit?

_____

_____

_____

_____

27. How does the sound from the speaker compare with the sound from the guitar?

_____

_____

_____

_____

NOW WHAT DO YOU THINK?

Take a minute to review your answers to the ScienceLog questions at the beginning of the chapter. Have your answers changed? If necessary, revise your answers based on what you have learned since you began this chapter. Record your revisions in your ScienceLog.

▲▲ **CHAPTER 19**

CHAPTER

20 VOCABULARY & NOTES WORKSHEET

The Energy of Waves

By studying the Vocabulary and Notes listed for each section below, you can gain a better understanding of this chapter.

SECTION 1

Vocabulary

In your own words, write a definition for each of the following terms in the space provided.

1. wave _____

_____

_____

2. medium _____

_____

3. transverse wave _____

_____

_____

_____

4. longitudinal wave _____

_____

_____

_____

Notes

Read the following section highlights. Then, in your own words, write the highlights in your ScienceLog.

• A wave is a disturbance that transmits energy.

• A medium is a substance through which a wave can travel. The particles of a medium do not travel with the wave.

• Waves that require a medium are called mechanical waves. Waves that do not require a medium are called electromagnetic waves.

• Particles in a transverse wave vibrate perpendicular to the direction the wave travels.

• Particles in a longitudinal wave vibrate back and forth in the same direction that the wave travels.

• Transverse and longitudinal waves can combine to form surface waves.

SECTION 2

Vocabulary

In your own words, write a definition for each of the following terms in the space provided.

1. amplitude _____

_____

_____

_____

2. wavelength _____

_____

_____

_____

_____

3. frequency _____

_____

_____

_____

4. wave speed _____

_____

_____

_____

Notes

Read the following section highlights. Then, in your own words, write the highlights in your ScienceLog.

- Amplitude is the maximum distance the particles in a wave vibrate from their rest position. Large-amplitude waves carry more energy than small-amplitude waves.
- Wavelength is the distance between two adjacent crests (or compressions) of a wave.
- Frequency is the number of waves that pass a given point in a given amount of time. High-frequency waves carry more energy than low-frequency waves.
- Wave speed is the speed at which a wave travels. Wave speed can be calculated by multiplying the wavelength by the wave's frequency.

SECTION 3

Vocabulary

In your own words, write a definition for each of the following terms in the space provided.

1. reflection _____

_____

2. refraction _____

_____

3. diffraction _____

_____

4. interference _____

_____

5. standing wave _____

_____

6. resonance _____

_____

Notes

Read the following section highlights. Then, in your own words, write the highlights in your ScienceLog.

- Waves bounce back after striking a barrier during reflection.
- Refraction is the bending of a wave when it passes at an angle from one medium to another.
- Waves bend around barriers or through openings during diffraction. The amount of diffraction depends on the wavelength of the waves and the size of the barrier or opening.
- The result of two or more waves overlapping is called interference.
- Amplitude increases during constructive interference and decreases during destructive interference.
- Standing waves are waves in which portions of the wave do not move and other portions move with a large amplitude.
- Resonance occurs when a vibrating object causes another object to vibrate at one of its resonant frequencies.

CHAPTER

20   CHAPTER REVIEW WORKSHEET

The Energy of Waves

USING VOCABULARY

For each pair of terms, explain the difference in their meanings.

1. longitudinal wave/transverse wave

_____

_____

_____

_____

2. frequency/wave speed _____

_____

_____

_____

3. wavelength/amplitude _____

_____

_____

_____

_____

4. reflection/refraction _____

_____

_____

_____

5. constructive interference/destructive interference _____

_____

_____

_____

_____

▲ **CHAPTER 20**

UNDERSTANDING CONCEPTS

Multiple Choice

6. As the wavelength increases, the frequency
   a. decreases.
   b. increases.
   c. remains the same.
   d. increases, then decreases.

7. Which wave interaction explains why sound waves can be heard around corners?
   a. reflection
   b. refraction
   c. diffraction
   d. interference

8. Refraction occurs when a wave enters a new medium at an angle because
   a. the frequency changes.
   b. the amplitude changes.
   c. the wave speed changes.
   d. None of the above

9. The speed of a wave with a frequency of 2 Hz (2/s), an amplitude of 3 m, and a wavelength of 10 m is
   a. 0.2 m/s.
   b. 5 m/s.
   c. 12 m/s.
   d. 20 m/s.

10. Waves transfer
   a. matter.
   b. energy.
   c. particles.
   d. water.

11. A wave that is a combination of longitudinal and transverse waves is a
   a. sound wave.
   b. light wave.
   c. rope wave.
   d. surface wave.

12. The wave property that is related to the height of a wave is the
   a. wavelength.
   b. amplitude.
   c. frequency.
   d. wave speed.

13. During constructive interference,
   a. the amplitude increases.
   b. the frequency decreases.
   c. the wave speed increases.
   d. All of the above

14. Waves that don't require a medium are
   a. longitudinal waves.
   b. electromagnetic waves.
   c. surface waves.
   d. mechanical waves.

_____

The Energy of Waves, continued

_____

Short Answer

15. Draw a transverse and a longitudinal wave. Label a crest, a trough, a compression, a rarefaction, and wavelengths. Also label the amplitude on the transverse wave.

16. What is the relationship between frequency, wave speed, and wavelength?

_____

_____

_____

_____

_____

17. Explain how two waves can cancel each other out.

_____

_____

_____

_____

_____

CHAPTER 20

CONCEPT MAPPING

18. Use the following terms to create a concept map: *wave, refraction, transverse wave, longitudinal wave, wavelength, wave speed, diffraction.*

CRITICAL THINKING AND PROBLEM SOLVING

19. After you set up stereo speakers in your school's music room, you notice that in certain areas of the room the sound from the speakers is very loud and in other areas the sound is very soft. Explain how interference causes this.

_____

_____

_____

_____

_____

_____

_____

The Energy of Waves, continued

20. You have lost the paddles for the canoe you rented, and the canoe has drifted to the center of the pond. You need to get the canoe back to shore, but you do not want to get wet by swimming in the pond. Your friend on the shore wants to throw rocks behind the canoe to create waves that will push the canoe toward shore. Will this solution work? Why or why not?

_____

_____

_____

_____

21. Some opera singers have voices so powerful they can break crystal glasses! To do this, they sing one note very loudly and hold it for a long time. The walls of the glass move back and forth until the glass shatters. Explain how this happens in terms of resonance.

_____

_____

_____

_____

_____

MATH IN SCIENCE

22. A fisherman in a rowboat notices that one wave crest passes his fishing line every 5 seconds. He estimates the distance between the crests to be 2 m and estimates the crests of the waves to be 0.4 m above the troughs. Using these data, determine the amplitude and wave speed of the waves. Remember that wave speed is calculated with the formula $v = \lambda \times f$.

_____

_____

_____

_____

_____

▲ CHAPTER 20

INTERPRETING GRAPHICS

23. Rank the waves below from highest energy to lowest energy, and explain your reasoning.

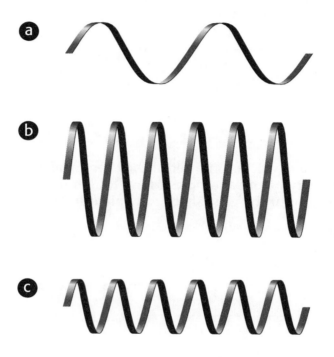

_____

_____

_____

_____

_____

_____

NOW WHAT DO YOU THINK?

Take a minute to review your answers to the ScienceLog questions at the beginning of this chapter. Have your answers changed? If necessary, revise your answers based on what you have learned since you began this chapter. Record your revisions in your ScienceLog.

The Nature of Sound

By studying the Vocabulary and Notes listed for each section below, you can gain a better understanding of this chapter.

SECTION 1

Vocabulary

In your own words, write a definition for the following terms in the space provided.

1. wave _____

_____

_____

2. medium _____

_____

_____

3. outer ear _____

_____

_____

4. middle ear _____

_____

_____

5. inner ear _____

_____

_____

Notes

Read the following section highlights. Then, in your own words, write the highlights in your ScienceLog.

• All sounds are created by vibrations and travel as longitudinal waves.

• Sound waves require a medium through which to travel.

• Sound waves travel in all directions away from their source.

• The sounds you hear are converted into electrical impulses by your ears and then sent to your brain for interpretation.

• Exposure to loud sounds can cause hearing loss and tinnitus.

SECTION 2

Vocabulary

In your own words, write a definition for the following terms in the space provided.

1. pitch _____

_____

_____

2. infrasonic _____

_____

_____

3. ultrasonic _____

_____

_____

4. Doppler effect _____

_____

_____

_____

5. loudness _____

_____

_____

6. decibel _____

_____

_____

Notes

Read the following section highlights. Then, in your own words, write the highlights in your ScienceLog.

• The speed of sound depends on the medium through which the sound is traveling. Changes in temperature of the medium can affect the speed of sound.

• The pitch of a sound depends on frequency. High-frequency sounds are high-pitched, and low-frequency sounds are low-pitched.

• Humans can hear sounds with frequencies between 20 Hz and 20,000 Hz.

The Nature of Sound, continued

- The Doppler effect is the apparent change in frequency of a sound caused by the motion of either the listener or the source of the sound.
- The loudness of a sound increases as the amplitude increases. Loudness is expressed in decibels.
- An oscilloscope can be used to "see" sounds.

SECTION 3

Vocabulary

In your own words, write a definition for the following terms in the space provided.

1. reflection _____

_____

_____

2. echo _____

_____

_____

3. echolocation _____

_____

_____

4. interference _____

_____

_____

5. sonic boom _____

_____

_____

_____

6. standing wave _____

_____

_____

7. resonance _____

_____

_____

_____

8. diffraction _____

_____

_____

Notes

Read the following section highlights. Then, in your own words, write the highlights in your ScienceLog.

• Echoes are reflected sound waves.

• Some animals use echolocation to find food or navigate around objects. Sonar and ultrasonography are types of echolocation.

• Sound barriers and shock waves are created by interference. You hear a sonic boom when a shock wave reaches your ears.

• Standing waves form at an object's resonant frequencies.

• Resonance occurs when a vibrating object causes a second object to vibrate at one of its resonant frequencies.

• The bending of sound waves around barriers or through openings is called diffraction. The amount of diffraction depends on the wavelength of the waves as well as the size of the opening.

SECTION 4

Vocabulary

In your own words, write a definition for the following terms in the space provided.

1. sound quality _____

_____

2. noise _____

_____

Notes

Read the following section highlights. Then, in your own words, write the highlights in your ScienceLog.

• Different instruments have different sound qualities.

• The three families of instruments are strings, winds, and percussion.

• The sound quality of noise is not pleasing because it is a random mix of frequencies.

CHAPTER

21 **CHAPTER REVIEW WORKSHEET**

The Nature of Sound

USING VOCABULARY

To complete the following sentences, choose the correct term from each pair of terms listed below, and write the term in the space provided.

1. Humans cannot hear _____ waves because their frequencies are above the range of human hearing. (infrasonic or ultrasonic)

2. In the _____, vibrations are converted to electrical signals for the brain to interpret. (middle ear or inner ear)

3. The _____ of a sound wave depends on its amplitude. (loudness or pitch)

4. Reflected sound waves are called _____ . (echoes or noise)

5. Two different instruments playing the same note sound different because of _____ . (echolocation or sound quality)

Multiple Choice

6. If a fire engine is traveling toward you, the Doppler effect will cause the siren to sound
   a. higher.
   b. lower.
   c. louder.
   d. softer.

7. The wave interaction most important for echolocation is
   a. reflection.
   b. interference.
   c. diffraction.
   d. resonance.

8. If two sound waves interfere constructively, you will hear
   a. a high-pitched sound.
   b. a softer sound.
   c. a louder sound.
   d. no change in sound.

9. You will hear a sonic boom when
   a. an object breaks the sound barrier.
   b. an object travels at supersonic speeds.
   c. a shock wave reaches your ears.
   d. the speed of sound is 290 m/s.

10. Instruments that produce sound when struck belong to which family?
   a. strings
   b. winds
   c. percussion
   d. None of the above

11. Resonance can occur when an object vibrates at another object's
   a. resonant frequency.
   b. fundamental frequency.
   c. second overtone frequency.
   d. All of the above

The Nature of Sound, continued

---

12. The amount of diffraction that a sound wave undergoes depends on

    a. the frequency of the wave.

    b. the amplitude of the wave.

    c. the size of the barrier.

    d. Both (a) and (c)

13. A technological device that can be used to "see" sound waves is a(n)

    a. oscilloscope.   **c.** transducer.

    b. sonar.   **d.** amplifier.

Short Answer

14. Describe how the Doppler effect helps a beluga whale determine whether a fish is moving away from it or toward it.

_____

_____

_____

_____

_____

_____

15. How is interference involved in forming a shock wave?

_____

_____

_____

_____

16. Briefly describe how the three parts of the ear work.

_____

_____

_____

_____

_____

_____

_____

CONCEPT MAPPING

17. Use the following terms to create a concept map: *sound, sound waves, hertz, pitch, loudness, decibels, frequency, amplitude.*

The Nature of Sound, continued

18. An anechoic chamber is a room where there is almost no reflection of sound waves. Anechoic chambers are often used to test sound equipment, such as stereos. The walls of such chambers are usually covered with foam triangles. Explain why this design eliminates echoes in the room.

_____

_____

_____

_____

19. Suppose you are sitting in the passenger seat of a parked car. You hear sounds coming from the stereo of another car parked on the opposite side of the street. You can easily hear the low-pitched bass sounds but cannot hear any high-pitched sounds coming from the parked car. Explain why you think this happens.

_____

_____

_____

_____

20. After working in a factory for a month, a man you know complains about a ringing in his ears. What might be wrong with him? What do you think may have caused his problem? What can you suggest to him to prevent further hearing loss?

_____

_____

_____

_____

_____

MATH IN SCIENCE

21. How far does sound travel in 4 seconds through water at 20°C and glass at 20°C? Refer to the chart on page 539 for the speed of sound in different media.

_____

_____

_____

INTERPRETING GRAPHICS

Use the oscilloscope screens below to answer the following questions:

22. Which sound is probably noise? _____

23. Which represents the softest sound? _____

24. Which represents the sound with the lowest pitch? _____

25. Which two sounds were produced by the same instrument? _____

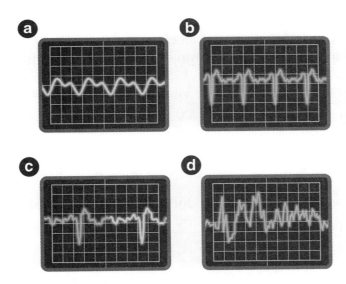

NOW WHAT DO YOU THINK?

Take a minute to review your answers to the ScienceLog questions at the beginning of the chapter. Have your answers changed? If necessary, revise your answers based on what you have learned since you began this chapter. Record your revisions in your ScienceLog.

VOCABULARY & NOTES WORKSHEET

The Nature of Light

By studying the Vocabulary and Notes listed for each section below, you can gain a better understanding of this chapter.

SECTION 1

Vocabulary

In your own words, write a definition for the following terms in the space provided.

1. electromagnetic wave _____

_____

_____

2. radiation _____

Notes

Read the following section highlights. Then, in your own words, write the highlights in your ScienceLog.

- Light is an electromagnetic (EM) wave. An electromagnetic wave is a wave that travels as vibrating electric and magnetic fields. EM waves require no medium through which to travel.

- Electromagnetic waves are produced by the vibration of electrically charged particles.

- The speed of light in a vacuum is 300,000,000 m/s.

SECTION 2

Vocabulary

In your own words, write a definition for the following term in the space provided.

1. electromagnetic spectrum _____

Notes

Read the following section highlights. Then, in your own words, write the highlights in your ScienceLog.

- All EM waves travel at the speed of light. EM waves differ only by wavelength and frequency.

- The entire range of EM waves is called the electromagnetic spectrum.

- Radio waves are most often used for communication.

- Microwaves are used for cooking and in radar.

- Infrared waves have shorter wavelengths and higher frequencies than microwaves. The absorption of infrared waves is felt as an increase in temperature.

- Visible light is the very narrow range of wavelengths that humans can see. Different wavelengths are seen as different colors.

The Nature of Light, continued

- Ultraviolet light is useful for killing bacteria and for producing vitamin D in the body, but overexposure can cause health problems.
- X rays and gamma rays are EM waves that are often used in medicine. Overexposure to these EM waves can damage or kill living cells.

SECTION 3

Vocabulary

In your own words, write a definition for the following terms in the space provided.

1. reflection _____

_____

2. law of reflection _____

_____

3. absorption _____

_____

4. scattering _____

_____

5. refraction _____

_____

6. diffraction _____

_____

7. interference _____

_____

Notes

Read the following section highlights. Then, in your own words, write the highlights in your ScienceLog.

- Two types of reflection are regular and diffuse reflection.
- Absorption and scattering cause light beams to become dimmer with distance.
- How much a light beam bends during refraction depends on the wavelength of the light.
- Light waves diffract more when traveling through a narrow opening.
- Interference of light waves can cause bright and dark bands.

_____
The Nature of Light, continued
_____

SECTION 4
Vocabulary
In your own words, write a definition for the following terms in the space provided.

1. transmission _____

2. transparent _____

_____

3. translucent _____

_____

_____

4. opaque _____

5. pigment _____

_____

Notes
Read the following section highlights. Then, in your own words, write the highlights in your ScienceLog.

• Objects are classified as transparent, translucent, or opaque depending on their ability to transmit light.

• Colors of opaque objects are determined by the color of light they reflect. White opaque objects reflect all colors and black opaque objects absorb all colors.

• Colors of transparent and translucent objects are determined by the color of light they transmit. All other colors are absorbed.

• White light is a mixture of all colors of light. The primary colors of light are red, blue, and green.

• Pigments give objects color. The primary pigments are magenta, cyan, and yellow.

CHAPTER

22 **CHAPTER REVIEW WORKSHEET**

The Nature of Light

USING VOCABULARY

To complete the following sentences, choose the correct term from each pair of terms listed below, and write the term in the space provided.

1. _____ is the transfer of energy by electromagnetic waves. (Radiation or Scattering)

2. This book is a(n) _____ object. (translucent or opaque)

3. _____ is a wave interaction that occurs when two or more waves overlap and combine. (Diffraction or Interference)

4. Light is a type of _____ . (electromagnetic wave or electromagnetic spectrum)

5. Light travels through an object during _____ . (absorption or transmission)

Multiple Choice

6. Electromagnetic waves transmit
  a. charges.
  b. fields.
  c. matter.
  d. energy.

7. Objects that transmit light easily are
  a. opaque.
  b. translucent.
  c. transparent.
  d. colored.

8. You can see yourself in a mirror because of
  a. absorption.
  b. scattering.
  c. regular reflection.
  d. diffuse reflection.

9. Shadows have blurry edges because of
  a. diffuse reflection.
  b. scattering.
  c. diffraction.
  d. refraction.

10. Microwaves are often used for
  a. cooking.
  b. broadcasting AM radio.
  c. cancer treatment.
  d. All of the above

11. What color of light is produced when red light is added to green light?
  a. cyan
  b. blue
  c. yellow
  d. white

12. Prisms produce rainbows through
  a. reflection.
  b. refraction.
  c. diffraction.
  d. interference.

The Nature of Light, continued

13. Which type of electromagnetic wave travels the fastest in a vacuum?
    a. radio waves
    b. visible light
    c. gamma rays
    d. They all travel at the same speed.

14. Electromagnetic waves are made of
    a. vibrating particles.
    b. vibrating charged particles.
    c. vibrating electric and magnetic fields.
    d. electricity and magnetism.

Short Answer

15. Name two ways EM waves differ from one another.

    _____

    _____

    _____

16. Describe how an electromagnetic wave is produced.

    _____

    _____

    _____

    _____

    _____

17. Why is it difficult to see through glass that has frost on it?

    _____

    _____

    _____

The Nature of Light, continued

CONCEPT MAPPING

18. Use the following terms to create a concept map: *light, matter, reflection, absorption, scattering, transmission.*

CHAPTER 22

The Nature of Light, continued

---

19. A tern is a type of bird that dives underwater to catch fish. When a young tern begins learning to catch fish, it is rarely successful. The tern has to learn that when a fish appears to be in a certain place underwater, the fish is actually in a slightly different place. Explain why the tern sees the fish in the wrong place.

_____

_____

_____

_____

_____

20. Radio waves and gamma rays are both types of electromagnetic waves. Exposure to radio waves does not harm the human body, whereas exposure to gamma rays can be extremely dangerous. What is the difference between these types of EM waves? Why are gamma rays more dangerous?

_____

_____

_____

_____

_____

21. If you look around a parking lot during the summer, you might notice sun shades set up in the windshields of cars. Explain how the sun shades help keep the inside of a car cool.

_____

_____

_____

_____

_____

The Nature of Light, continued

MATH IN SCIENCE

22. Calculate the time it takes for light from the sun to reach Mercury. Mercury is 54,900,000,000 m away from the sun.

_____

_____

_____

_____

_____

_____

_____

INTERPRETING GRAPHICS

23. Refer to question 23 on page 589 of your text. Each of the pictures shows the effects of a wave interaction of light. Identify the interaction involved.

_____

_____

_____

NOW WHAT DO YOU THINK?

Take a minute to review your answers to the ScienceLog questions at the beginning of the chapter. Have your answers changed? If necessary, revise your answers based on what you have learned since you began this chapter. Record your revisions in your ScienceLog.

CHAPTER

23 **VOCABULARY & NOTES WORKSHEET**

Light and Our World

By studying the Vocabulary and Notes listed for each section below, you can gain a better understanding of this chapter.

SECTION 1

Vocabulary

In your own words, write a definition for the following terms in the space provided.

1. luminous _____

_____

2. illuminated _____

_____

3. incandescent light _____

4. fluorescent light _____

_____

5. neon light _____

_____

6. vapor light _____

_____

Notes

Read the following section highlights. Then, in your own words, write the highlights in your ScienceLog.

- You see objects either because they are luminous (produce their own light) or because they are illuminated (reflect light).
- Light produced by hot objects is incandescent light. Ordinary light bulbs are a common source of incandescent light.
- Fluorescent light is visible light emitted by a particle when it absorbs ultraviolet light. Little energy is wasted by fluorescent light bulbs.
- Neon light results from an electric current in certain gases.
- Vapor light is produced when electrons combine with gaseous metal atoms.

Light and Our World, continued

SECTION 2

Vocabulary

In your own words, write a definition for the following terms in the space provided.

1. plane mirror _____

2. concave mirror _____

3. focal point _____

_____

_____

4. convex mirror _____

_____

5. lens _____

_____

_____

6. convex lens _____

7. concave lens _____

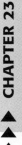

Notes

- Rays are arrows that show the path and direction of a single light wave. Ray diagrams can be used to determine where images are formed by mirrors and lenses.
- Plane mirrors produce virtual images that are the same size as the objects. These images are reversed left to right.
- Concave mirrors can produce real images and virtual images. They can also be used to produce a powerful light beam.
- Convex mirrors produce only virtual images.
- Convex lenses can produce real images and virtual images. A magnifying glass is an example of a convex lens.
- Concave lenses produce only virtual images.

SECTION 3

Vocabulary

In your own words, write a definition for the following terms in the space provided.

1. cornea _____

2. pupil _____

3. iris _____

4. retina _____

Notes

Read the following section highlights. Then, in your own words, write the highlights in your ScienceLog.

• Your eye has several parts, such as the cornea, the pupil, the iris, the lens, and the retina.

• Nearsightedness and farsightedness occur when light is not focused on the retina. Both problems can be corrected with glasses or contact lenses.

• Color deficiency is a genetic condition in which cones in the retina are given the wrong instructions. Color deficiency cannot be corrected.

SECTION 4

Vocabulary

In your own words, write definition for the following terms in the space provided.

1. laser _____

2. hologram _____

_____

Notes

Read the following section highlights. Then, in your own words, write the highlights in your ScienceLog.

• Optical instruments, such as cameras, telescopes, and microscopes, are devices that use mirrors and lenses to help people make observations.

• Lasers are devices that produce intense, coherent light of only one wavelength and color. Lasers produce light by a process called stimulated emission.

• Optical fibers can transmit light over long distances because of total internal reflection.

• Polarized light contains light waves that vibrate in only one direction.

CHAPTER

23 CHAPTER REVIEW WORKSHEET

Light and Our World

USING VOCABULARY

To complete the following sentences, choose the correct term from each pair of terms listed below, and write the term in the space provided.

1. _____ is commonly used in homes and produces a lot of thermal energy. (Incandescent light or Fluorescent light)

2. A _____ is curved inward, like the inside of a spoon. (convex mirror or concave mirror)

3. You can see an object when light is focused on the

   _____ of your eye. (pupil or retina)

4. A _____ is a device that produces coherent, intense light of only one color. (laser or lens)

5. You can see this book because it is a(n) _____ object. (luminous or illuminated)

UNDERSTANDING CONCEPTS

Multiple Choice

6. When you look at yourself in a plane mirror, you see a
   a. real image behind the mirror.
   b. real image on the surface of the mirror.
   c. virtual image that appears to be behind the mirror.
   d. virtual image that appears to be in front of the mirror.

7. A vision problem that occurs when light is focused in front of the retina is
   a. nearsightedness.
   b. farsightedness.
   c. color deficiency.
   d. None of the above

8. Which part of the eye refracts light?
   a. iris
   b. cornea
   c. lens
   d. both (b) and (c)

9. Visible light produced when electrons combine with gaseous metal atoms is
   a. incandescent light.
   b. fluorescent light.
   c. neon light.
   d. vapor light.

10. You see less of a glare when you wear certain sunglasses because the lenses
    a. produce total internal reflection.
    b. create holograms.
    c. produce coherent light.
    d. polarize light.

11. What kind of mirrors provide images of large areas and are used for security?
    a. plane mirrors
    b. concave mirrors
    c. convex mirrors
    d. All of the above

12. A simple refracting telescope has
  a. a convex lens and a concave lens.
  b. a concave mirror and a convex lens.
  c. two convex lenses.
  d. two concave lenses.

13. Light waves in a laser beam interact and act as one wave. This light is called
  a. red.   **c.** coherent.
  b. white.   **d.** emitted.

Short Answer

14. What type of lens should be prescribed for a person who cannot focus on nearby objects? Explain.

_____

_____

_____

_____

15. How is a hologram different from a photograph?

_____

_____

_____

_____

16. Why might a scientist at the North Pole need polarizing sunglasses?

_____

_____

_____

_____

Light and Our World, continued

CONCEPT MAPPING

17. Use the following terms to create a concept map: *lenses, telescopes, cameras, real images, virtual images, optical instruments.*

▲ CHAPTER 23

18. Stoplights are usually mounted so that the red light is on the top and the green light is on the bottom. Explain why it is important for a person who has red-green color deficiency to know this arrangement.

_____

_____

_____

_____

_____

_____

19. Some companies are producing fluorescent light bulbs that will fit into sockets on lamps designed for incandescent light bulbs. Although fluorescent bulbs are more expensive, the companies hope that people will use them because they are better for the environment. Explain why fluorescent light bulbs are better for the environment than incandescent light bulbs.

_____

_____

_____

_____

20. Imagine you are given a small device that produces a beam of red light. You want to find out if the device is producing laser light or if it is just a red flashlight. To do this, you point the beam of light against a wall across the room. What would you expect to see if the device is producing laser light? Explain.

_____

_____

_____

21. Examine the ray diagrams below, and identify the type of mirror or lens that is being used and the kind of image that is being formed.

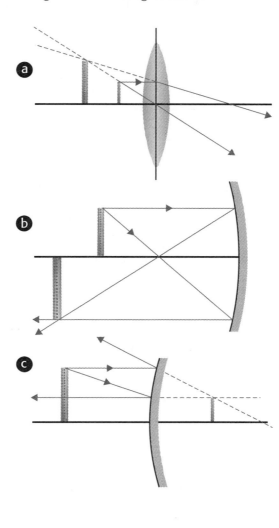

_____

_____

_____

_____

_____

_____

NOW WHAT DO YOU THINK?

Take a minute to review your answers to the ScienceLog questions at the beginning of the chapter. Have your answers changed? If necessary, revise your answers based on what you have learned since you began this chapter. Record your revisions in your ScienceLog.